Lecture Notes in Computer Science   13122

More information about this subseries at http://www.springer.com/series/7407

João F. Ferreira · Alexandra Mendes ·
Claudio Menghi (Eds.)

Formal Methods Teaching

4th International Workshop and Tutorial, FMTea 2021
Virtual Event, November 21, 2021
Proceedings

Editors
João F. Ferreira (ID)
INESC-ID & IST
University of Lisbon
Lisbon, Portugal

Alexandra Mendes (ID)
INESC TEC and University of Beira Interior
Covilhã, Portugal

Claudio Menghi (ID)
McMaster University
Hamilton, ON, Canada

ISSN 0302-9743   ISSN 1611-3349   (electronic)
Lecture Notes in Computer Science
ISBN 978-3-030-91549-0   ISBN 978-3-030-91550-6   (eBook)
https://doi.org/10.1007/978-3-030-91550-6

LNCS Sublibrary: SL1 – Theoretical Computer Science and General Issues

This Springer imprint is published by the registered company Springer Nature Switzerland AG
The registered company address is: Gewerbestrasse 11, 6330 Cham, Switzerland

Preface

Formal Methods provide software engineering with tools and techniques for rigorously reasoning about the correctness of systems. While in recent years formal methods are increasingly being used in industry, university curricula are not adapting at the same pace. Some existing formal methods courses interest and challenge students, whereas others fail to ignite student motivation. It is thus important to develop, share, and discuss approaches to effectively teach formal methods to the next generation. This discussion is now more important than ever due to the challenges and opportunities that arose from the pandemic, which forced many educators to adapt and deliver their teaching online. The exchange of ideas is critical to making these new online approaches a success and having a greater reach.

FMTea 2021 (Formal Methods Teaching Workshop and Tutorial) was a combined workshop and tutorial on teaching formal methods held on November 21, 2021, as part of Formal Methods, FM 2021.

The workshop received 12 submissions: eight were accepted as full papers and two as short papers. The review process was single-blind. All the papers were reviewed by at least two members of the Program Committee.

The workshop was organized in five sessions. The authors presented their papers and contributed to a lively discussion of their topic and its alternatives together with their peers and the audience.

The technical program also included three invited talks:

- Tobias Nipkow (Technical University Munich, Germany) on his experiences with a new course on interactive theorem proving that relies on the Isabelle proof assistant, taught at the Technical University of Munich.
- Jeremy Avigad (Carnegie Mellon University, USA) on his experiences in teaching a second-year undergraduate CS course based on Lean 4.
- Laura Kovács (Vienna University of Technology, Austria) on her experiences with online teaching for two master courses in the Logic and Computation curricula of the TU Wien.

We would like to thank all of our invited speakers for agreeing to present at our workshop and for the exciting and inspiring ideas they brought to it.

FMTea 2021 would not have been possible without the support of the FM 2021 general chair, Huimin Lin, the workshop and tutorial chairs, Carlo A. Furia, Lijun Zhang, Luigia Petre and Tim A.C. Willemse, and the numerous people involved in the local organization of FM 2021. We are grateful for their enthusiasm and dedication. We would also like to thank the Program Committee for their opinions on the papers we received, and of course the authors for sharing their innovative teaching practices.

Finally, we acknowledge EasyChair, which supported us in the submission and reviewing process, as well as in generating these proceedings and the FMTea 2021 program.

October 2021   João F. Ferreira
                                                                               Alexandra Mendes
                                                                                Claudio Menghi

Organization

Program Committee Chairs

João F. Ferreira   INESC-ID & IST, University of Lisbon, Portugal
Alexandra Mendes   INESC TEC and University of Beira Interior, Portugal
Claudio Menghi   McMaster University, Canada

Program Committee

Sandrine Blazy   University of Rennes 1, France
Brijesh Dongol   University of Surrey, UK
Catherine Dubois   ENSIIE, France
Rustan Leino   Amazon Web Services, USA
José N. Oliveira   University of Minho, Portugal
Luigia Petre   Åbo Akademi University, Finland
Kristin Rozier   Iowa State University, USA
Pierluigi San Pietro   Politecnico di Milano, Italy
Emil Sekerinski   McMaster University, Canada
Graeme Smith   The University of Queensland, Australia

Invited Talks

Teaching Logic and Mechanized Reasoning with Lean 4

Jeremy Avigad

Carnegie Mellon University, Department of Philosophy, Baker Hall 161,
Pittsburgh, PA 15213

Abstract. This semester, I am co-teaching a course on logic and mechanized reasoning with Marijn Heule, addressed to second-year undergraduate students in computer science. The course is based on the Lean 4 programming language and proof assistant. In this talk, I will introduce Lean 4, describe the course, and report on our experiences.

Automating Teaching Efforts for Deductive Verification

Laura Kovács

TU Wien, Austria

Abstract. Amid the COVID-19 pandemic, higher education has moved to distance teaching. While online lecturing was relatively fast to implement via webinars, recordings, streaming and online communication channels, coming up with best practices to assess course performance was far from trivial. In this talk, I will present the setting we used for online teaching and online examination for two master courses in the Logic and Computation curricula of the TU Wien. Namely, I will describe our courses of "Automated Deduction" and "Formal Methods in Computer Science", present the SAT/SMT/first-order reasoning framework we use in these courses and their link to deductive verification. I will also detail the framework we designed for automating the generation of exam sheets for these course; for this process of generating exam sheets, we actually used the reasoning/verification tools Z3, Vampire and Absynth.

Teaching Data Structures and Algorithms with a Proof Assistant

Tobias Nipkow

Technische Universität München, Munich, Germany

Abstract. I will report on a new course "Verified Functional Data Structures and Algorithms" taught at the Technical University of Munich. The course first introduces students to interactive theorem proving with the Isabelle proof assistant. Then it covers a range of standard data structures, in particular search trees and priority queues. It is shown how to express these data structures functionally and how to reason about their correctness and running time in Isabelle.

Contents

Introducing Formal Methods to First-Year Students in Three
Intensive Weeks   1
   Luca Aceto and Anna Ingólfsdóttir

Online Teaching of Verification of C Programs in Applied Computer
Science   18
   Matthias Güdemann

A Proposal for a Framework to Accompany Formal Methods Learning
Tools (Short Paper).   35
   Norbert Hundeshagen and Martin Lange

Increasing Engagement with Interactive Visualization: Formal Methods as
Serious Games   43
   Eduard Kamburjan and Lukas Grätz

Increasing Student Self-Reliance and Engagement in Model-Checking
Courses   60
   Philipp Körner and Sebastian Krings

Teaching Formal Methods to Software Engineers through Collaborative
Learning (Short Paper).   75
   Livia Lestingi

Lessons of Formal Program Design in Dafny   84
   Ran Ettinger

Teaching Correctness-by-Construction and Post-hoc Verification – The
Online Experience.   101
   Tobias Runge, Tabea Bordis, Thomas Thüm, and Ina Schaefer

Using Isabelle in Two Courses on Logic and Automated Reasoning   117
   Jørgen Villadsen and Frederik Krogsdal Jacobsen

Introducing Formal Methods to Students Who Hate Maths and Struggle
with Programming.   133
   Nisansala Yatapanage

Author Index   147

Contents

Introducing Formal Methods to First-Year Students in Three Intensive Weeks

Luca Aceto[1,2]([✉]) [iD] and Anna Ingólfsdóttir[1] [iD]

[1] ICE-TCS, Department of Computer Science, Reykjavik University,
Reykjavik, Iceland
{luca,annai}@ru.is
[2] Gran Sasso Science Institute, L'Aquila, Italy

Abstract. This paper presents a crash course whose goal is to introduce modelling and verification using model-checking technology to mostly first- and second-year bachelor students at the Department of Computer Science at Reykjavik University. The course is student driven, project based and fosters independent learning in the student body. During the course, students tackle a number of non-trivial modelling and verification tasks using the model checker Uppaal, while also practising 'soft skills' such as their communication skills, as well as their ability to work independently and as members of a team.

Keywords: Formal methods education · Modelling and verification · Timed automata · Uppaal

1 Introduction

This article describes the Real-Time Models course (henceforth abbreviated to REMO), its context and its structure. REMO is a three-week, intensive, introductory course on 'applied formal methods' we designed and have taught, individually or together, at Reykjavik University every year since the spring semester 2013.

The main goal of the REMO course is to give bachelor students at the Department of Computer Science at Reykjavik University a hands-on introduction to modelling and verification in a student-driven, project-based setting. During the course, students tackle a number of non-trivial modelling and verification tasks using the model checker Uppaal, an integrated tool environment for modelling, validation and verification of real-time systems modelled as networks of timed automata, extended with data types. As part and parcel of the course, students also hone their presentation and writing skills, as well as their ability to work

This work has been partly funded by the projects "Open Problems in the Equational Logic of Processes (OPEL)" (grant no. 196050) and "MoVeMnt: Mode(l)s of Verification and Monitorability" (grant no. 217987) of the Icelandic Research Fund.

J. F. Ferreira et al. (Eds.): FMTea 2021, LNCS 13122, pp. 1–17, 2021.
https://doi.org/10.1007/978-3-030-91550-6_1

independently and as members of a team. A number of students must follow the course in the first year of their studies, while others take it in their second or third year as an elective. This means that the course cannot assume any previous knowledge on the part of the students apart from programming and discrete mathematics. For many of the students following the course, including those in our Software Engineering BSc programme, the REMO course will provide the only opportunity to become acquainted with formal methods during their studies. In our, admittedly biased, opinion, this state of affairs is undesirable because we believe that every student graduating with a bachelor degree in a Computer-Science-related subject should have some working knowledge of 'applied formal methods'[1]. Indeed, our graduates will be the next generation of designers and developers of computing systems that will permeate the daily operations of our future society even more than they do today. This population of devices is already embedded in the fabric of our homes, shops, vehicles, farms and some even in our bodies. They help us command, control, communicate, do business, travel and entertain ourselves.

In light of the increasing complexity of computing systems, and of the fact that they control important, when not altogether safety critical, operations, we think that our students should realise that it is important to adopt high standards of quality in their development and validation, and that a key scientific challenge in computer science is to design and develop computing systems that do what they were designed to do, and do so reliably.

Our goal in the REMO course is to expose early-career bachelor students to the use of model-based approaches in the design and validation of computing systems. As discussed in Sects. 2 and 3 of this article, we do so in an intensive three-week course, during which the students use the model checker Uppaal and its underlying modelling formalism to model and analyse algorithms, games, scheduling problems and other fun scenarios with relevance to Computer Science. During the course, we focus on the use of Uppaal and introduce only the bare minimum of the underlying theoretical foundations the students really need in order to make a principled use of the tool in addressing the challenges we pose them. Moreover, we limit ourselves to presenting the features of the modelling formalism and of the tool that are relevant for the modelling and verification tasks tackled by the students. Our hope is that, after having followed this course, students will realise that the use of formal methods can have impact on the practice of the development of computing systems in a world that increasingly depends on the quality of software-controlled devices. Moreover, in our experience, students who take the course then go on to follow our master-level Modelling and Verification course or are well-equipped to take similar courses at other institutions. In addition, they serve as ambassadors for model checking technology, and formal methods in general, by enticing other students to

---

[1] We note, in passing, that in Iceland many of our students find well-paid jobs even before graduating with a bachelor degree and do not pursue master-level studies. To our mind, this phenomenon increases the importance of exposing them to formal methods during their bachelor studies.

follow the REMO course and by informing their co-workers of the usefulness of modelling and verification.

The rest of the paper describes the REMO course in more detail by presenting the context for course (Sect. 2), its goals and underlying pedagogical philosophy (Sect. 3), and its structure (Sect. 4). We also introduce the two 'pandemic editions' of the course and how we adapted a course designed for supervised, intensive work in class to an online setting (Sect. 5). We conclude the paper with a brief evaluation of the course based on the opinions we have received from the roughly 300 students who have followed the course since 2013, and with a discussion of some possible future steps that might increase its impact (Sect. 6).

2   Context for the Course

In each semester, teaching at Reykjavik University is divided into two distinct periods, each followed immediately by exams. The first part of the semester lasts for 12 weeks, during which students typically follow four six-ECTS courses concurrently[2]. The second part of the semester spans three weeks, which are devoted to one six-ECTS course that is taught in 'full-immersion mode'. During those three weeks, students are expected to engage in activities related to the single course they are following every working day for about eight hours per day.

Even though we have occasionally taught courses involving a substantial theoretical component during the three-week period ourselves, courses held at that time mostly have a project-based and practical component, including a focus on group work. To our mind, and based on our experience in teaching it since the spring of 2013, the REMO course fits the three-week period very well.

The course was originally designed for students in the three-year bachelor programme in Discrete Mathematics and Computer Science (DIMACS) at Reykjavik University, where it is a compulsory second-semester course. However, it is also available as an elective for students in the bachelor programmes in Computer Science and in Software Engineering. To put the students' knowledge in context, we remark that all students who enrol in the course have followed two courses in programming (Programming, Data Structures), one Computer Architecture course as well as one or two courses in Discrete Mathematics (which briefly introduce propositional logic, elementary graph theory, finite automata, grammars and regular expressions amongst many other topics). In addition, students in the DIMACS programme have taken two Calculus courses and a Linear Algebra course. Apart from testing their programs in the programming courses,

---

[2] The European Credit Transfer and Accumulation System (ECTS) is used within the European Higher Education Area to make studies and courses transparent and to allow students to transfer study credits between institutions, possibly located in different countries, in a seamless fashion. Depending on the country, one ECTS credit point corresponds to an average between 25 and 30 actual study hours. A bachelor-level degree course is equivalent to 180 ECTS. See https://ec.europa.eu/education/resources-and-tools/european-credit-transfer-and-accumulation-system-ects_en for more information.

none of the students taking the REMO course has any familiarity with the theory and software tools underlying modelling and verification of computing systems, the field of program correctness, and topics such as concurrency, model checking, temporal logics and real-time systems. Therefore, when 'teaching' the course, we can only rely on the students' programming experience and on their willingness to engage actively with novel and, to many of them, alien material.

3   Goals and Overall Philosophy of the Course

The main knowledge-related aims of the REMO course are

- to introduce students early in their bachelor studies to the basic ideas underlying modelling and verification of computing systems,
- to help them to develop an appreciation of the importance of those activities in the development of computing systems, and
- to make them aware of the fact that there is powerful and eminently usable tool support they can employ in their modelling and verification tasks.

Moreover, as part of the course, students develop an appreciation of the key role that models play in Computer Science, of quality criteria good models should possess, and of how models and model checkers can be used to synthesise control programs, plans and schedules satisfying a number of correctness and optimality criteria at the press of a button. In our experience, this last point is important, since students learn that, at times, it is best to describe computational tasks in a 'declarative' fashion and let our computational engines develop correct and 'optimal' algorithms for solving them on our behalf.

Since the course runs over three weeks and we want the students to be in a position to apply modelling and verification techniques already on the second day of the course, there is really room for presenting only *one* modelling formalism and *one* model checker based on it. Moreover, the course focuses solely on the application of model checking to a variety of problems and we eschew any mention of the underlying mathematical theory and algorithmics, apart from hinting at why the computational problems solved by the model checker are computationally hard. Our underlying philosophy in this course is that less is more; rather than drowning students in theoretical developments and tool features that they do not need in their modelling and verification challenges, we focus on introducing the bare necessities exactly when the students need them in a timely fashion. We trust that, having followed our introductory course, at a later stage in their studies, some of the students will be enticed to enrol in the master-level Modelling and Verification course we offer[3]—see [2] for a description of that course, which is based on the textbook [1].

---

[3] According to the data we have available, to date 26% of the student who followed the REMO course in the period 2013–2019 then went on to take the Modelling and Verification course. However, several of our students pursued master-level studies abroad. We expect that many of them took advanced courses on formal methods at foreign universities, but have no hard data to support this expectation.

The REMO course is centred on the seminal model of *(networks of) timed automata*, a graphical formalism for the description of real-time computing systems due to Rajeev Alur and David Dill [3]. During the course, students use the model to describe a variety of scenarios with relevance to Computer Science, and to analyse the behaviour of the systems they have modelled using the automatic verification tool Uppaal [5,6]. Uppaal is an integrated tool environment for the description, validation and verification of real-time systems modelled as networks of communicating timed automata, extended with data types.

In our course, we use the model checker Uppaal for a number of reasons. First, in our experience, students learn to use the basic features of the tool in a few hours and, after reading Frits Vaandrager's excellent introduction to the tool [17] and playing with the models accompanying that article, are ready to make and analyse their first models already by the start of the second day of the course. This opinion of ours is confirmed by Roelof Hamberg and Frits Vaandrager, amongst others, who have used the Uppaal model checker in an introductory course on operating systems for first-year Computer Science students at the Radboud University Nijmegen [13]. We think that the graphical nature of Uppaal models and the ease of use of the tool are crucial in an intensive course for students at the early stage of their bachelor studies, as these characteristics allow them to experience the usefulness of formal methods without having to understand modelling formalisms and tools with a steep learning curve. (For what it is worth, this opinion of ours is confirmed by the reports we have received from our students since the first edition of the course ran in 2013.) A second reason for using Uppaal in the course is that it is the model checking tool with which we have most familiarity, which ensures that our teaching assistants and we can provide timely answers to questions from the students. Moreover, having worked at Aalborg University with the people responsible for developing and maintaining the tool for many years, we can ask for their assistance regarding technical issues that may arise during the students' work. To our mind, prompt feedback on a variety of issues is crucial in a course that proceeds at a fast pace and lasts only three weeks.

Apart from the aforementioned goals related to the teaching of selected topics in formal methods, the REMO course also has the following pedagogical aims. During the course, students are made responsible for their own learning, and develop independence and peer-learning skills. One of the decisions we made in designing the course is that, during it, we do very little conventional lecturing. Moreover, the little lecturing we do is confined to the first morning of the course and to short sessions, during which we introduce increasingly sophisticated features in Uppaal when they may be of help for the students while they work in groups on a variety of modelling tasks.

The course is project based and student driven. One of our pedagogical tenets is that students should be prime movers in their own learning and that they can, and indeed do, challenge themselves when given the power to shape their own learning tasks. In order to entice them to do so, our course emphasises the playful aspects of modelling and verification activities, introducing important

Computer Science topics in fun and recreational settings. (See Sect. 4 for more details on the course structure.) Each student group reads the suggested material independently and solves the given modelling challenges at its own speed. Our role is to act as facilitators and to give students (hopefully helpful) hints when they have problems in their modelling work or face the state-explosion problem while verifying their models. A second important soft skill the students hone in the course is an ability to work both independently and as members of a group of four–five students. Last, but by no means least, students develop their technical-communication skills since they have to present their work both orally, as part of two conference sessions and at a final oral exam, and in writing, in the form of two project reports.

4   Structure of the Course

As mentioned above, the REMO course runs over a period of three weeks in the spring semester at Reykjavik University. (Interested readers may found some information on the 2021 edition of the course at http://www.icetcs.ru.is/remo-course/.) Each course day lasts roughly from 8:30 till 17:00. The mornings are mostly devoted to supervised independent work by the student groups. Students work independently in the afternoons, but we are available to answer their questions and to assist them in resolving issues they might encounter.

4.1   Week One: Warming up

The first week of the course serves as a warm-up period in which the students become acquainted with the context for the course, the Uppaal tool and the model of networks of timed automata it supports, and start working on a variety of modelling challenges. Nearly all the actual lecturing we do in the course is concentrated on the first morning of the first day of the course. During that course session, we introduce students to the context for the REMO course, presenting the correctness problem for computing systems as one of the key scientific challenges in Computer Science from its early days to today, and highlight various approaches to modelling and verification of computing systems, pointing to their applications in industry and to the development of trustworthy real-life applications with which students are familiar[4]. We also briefly introduce the model of networks of timed automata and the basic aspects of the query language supported by Uppaal, initially ignoring timing-related features. We typically do so in two steps. First, we discuss a version of the 'small university' system described in [1]. The goal of that very simple example is to highlight how a system can be described using automata running in parallel and communicating via synchronous handshakes. We present a deadlock-free version of that system and a variation that can exhibit a deadlock, analysing their behaviour by hand with

---

[4] The slides we used for this introduction in the 2021 edition of the course are available at http://www.icetcs.ru.is/remo-course/remo-intro.pdf.

the students' help. Next, we discuss the actual Uppaal model of the classic Gossiping problem (see, for instance, [14]) presented by Vaandrager in [17]. The Gossiping model serves a number of purposes at this very early stage in the course. In particular, it allows us to present some of the data types supported by Uppaal and some of the key features students will employ in their models (such as the concepts of nondeterministic selection of values for variables, and the use of guards and updates on transitions), as well as the use of the Uppaal verifier and its diagnostic-trace option to find an optimal sequence of phone calls the agents can use to share all their secrets.

The model of the Gossiping problem also highlights early on in the course one of the messages that we want students to take home, namely that the model focuses on describing *what* each agent can do at any given time given its current 'state' rather than on *how* the agents can optimally achieve their goal. Once a faithful model has been built and the goal the system should achieve has been expressed as a suitable reachability query in the Uppaal specification language, finding the optimal scheduling of phone calls is best left to the computational engine of the tool. In our experience, even for students in the age of machine learning, moving from an algorithmic and procedural way of thinking to a declarative one feels like a Copernican revolution for many students, and is best stressed right away and repeatedly in the course. Moreover, the analysis of the Gossiping model using Uppaal gives students their first introduction to the state-explosion problem. We show them how the processing time used by the tool increases substantially with the number of agents in the system. We think that this fact of Computer Science life is also worth highlighting early on in the course, since students are used to getting fast response from their devices to just about any computational task they have faced so far. As the course develops, they will see that the choices they make in their modelling tasks can crucially affect the time it takes for their verification efforts to complete, and that they might have to show some patience in waiting for an answer to their queries.

At the end of the morning session on the first day of the course, students form groups, which typically consist of four or five students, and we encourage them to spend the afternoon reading Vaandrager's Uppaal tutorial [17], to watch a video describing the Uppaal tool produced by one of our PhD students and to familiarise themselves with Uppaal by examining the models accompanying Vaandrager's introductory article.

By the start of the second day of the course, most student groups are typically ready to start working on modelling and verification challenges. We suggest that they begin by analysing the *How much can you lose?*[5] and *How much can we reach?*[6] puzzles and a number of classic mutual-exclusion algorithms. These warm-up challenges show students how they can turn pseudo-code descriptions of algorithms into Uppaal models, and introduce them to some of the conceptual challenges and classic pitfalls in concurrent programming. In order to make it easier for them to build models efficiently, we devote some time to showing

---

[5] See Exercise 23 at http://people.cs.aau.dk/~kgl/ESV04/exercises/index.html.
[6] See Exercise 24 at http://people.cs.aau.dk/~kgl/ESV04/exercises/index.html.

students how to express programming constructs such as loops and conditionals in Uppaal and how, using state transitions in Uppaal models, they can describe easily which operations are atomic and which are not, which is a key issue in concurrent programming. Moreover, using the Uppaal simulator and verifier, students can explore the effect that different atomicity assumptions have on the behaviour of apparently simple concurrent programs, leading to results that, at least at first sight, appear counter-intuitive.

A key (and fun) modelling challenge we pose the students early on during the first week of the course is to model the one- and two-dimensional solitaire games described in [4, Chap. 6][7] using Uppaal, and to employ the Uppaal verifier to check that the games can indeed be solved and to find solutions for them involving the least number of moves. This modelling exercise sets the stage for the two group projects on which the students work in the last two weeks of the course, and further reinforces the usefulness of declarative models.

As the first week of the course evolves, we introduce students to some of the timing-related features in Uppaal, always striving to focus on providing the least amount of information students need to use them properly in their models. In particular, we present the use of clocks in Uppaal models to express time-dependent system behaviour (with focus on guards and invariants), the peculiarities of variables of data type `clock`, the way in which time elapses in a network of timed automata and advanced features such as urgent channels as well as committed and urgent locations. Students employ these features right away in synthesising the control program for a coffee machine in the problem described at

http://people.cs.aau.dk/~kgl/ESV04/exercises/index.html#coffee.

The work done by the students during the first week of the course does *not* contribute to their final grade. We do so to allow them to explore the material independently and at their own speed, giving them enough time to become familiar with Uppaal, its underlying model and query language and to hone their modelling skills. We think that this decision of ours contributes to creating a positive atmosphere in the course in which students feel that they can learn by making mistakes, without affecting their final grades.

We devote the start of the course session on the last day of the first week of the course to a 'conference-like session', which we chair. During that session, each group of students delivers a short talk presenting their solution to one of the problems they tackled during the week to everyone involved in the course. Each presentation is followed by questions and comments from the audience, both on the quality of the modelling work and of the presentation. We stress to the students that giving good presentations based on their work is a skill they will need in their future careers, but one whose importance is not widely appreciated and that is, unfortunately, not practised sufficiently in many degree courses in Computer Science.

---

[7] See http://www.ru.is/faculty/luca/MV2011/solitaire.pdf for a scan of the relevant pages.

We close the first week of the course by presenting the first project for the course, so that students who want to start working on it can do so right away.

4.2 Week Two: First Project

The last two weeks in the course are mainly devoted to two group projects, which together account for 70% of the final grade for the course and form the basis for the final oral exam. In keeping with the 'recreational' atmosphere of the course, both projects entice students to explore challenging topics in formal methods in a 'serious-game setting'[8]. The projects we set our students change regularly, but the ones we describe in this section and in the subsequent one are two of our favourite ones and we have used them for the last three editions of the REMO course.

The project we have recently used for week two of the course is inspired by Vaandrager and Verbeek's article on designing vacuum-cleaning trajectories [18]. Working on it, apart from honing their modelling and verification skills, and learning about the connection between winning strategies in games and control software for a simple robotic-inspired application, students realise that automated support is needed to avoid (or at least reduce the number of) errors that human programmers make in developing even relatively simple systems.

Briefly, in their first project, students work with a vacuum-cleaning-robot problem from the book [19]. On pages 67–69 of that book[9], Wooldridge describes an example of a small robotic agent that will vacuum clean a room. In our version of the example, the room is a 3-by-3 grid and at any point the robot can move forward one step or turn clockwise 90°. The problem is to find a deterministic, memoryless strategy for the robot in which

1. its next action only depends on its current square and orientation (one of north, west, south, east), and
2. all squares are visited infinitely often.

Wooldridge gives a partial specification of such a strategy using a number of rules. Ignoring the actions of the robot having to do with sucking dirt and focussing only on the actions related to movement, the rules given by Wooldridge are:

- If $In(0,0)$ and $Facing(north)$ then $Do(forward)$.
- If $In(0,1)$ and $Facing(north)$ then $Do(forward)$.
- If $In(0,2)$ and $Facing(north)$ then $Do(turn)$.
- If $In(0,2)$ and $Facing(east)$ then $Do(forward)$.

---

[8] Other courses in formal methods employ recreational problems to engage students to good effect. By way of example, we limit ourselves to mentioning that Rozier has used magic-square, chess and Rubik's cube puzzles in an applied formal methods course offered to undergraduate and graduate students in Aerospace Engineering, Computer Science and Computer Engineering at Iowa State University [15].

[9] The relevant pages are available at http://icetcs.ru.is/fm-at-work/IntroductiontoM ultiAgentSystemsWooldridgepp67-69.pdf.

According to Wooldridge, 'similar rules can easily be generated that will get the agent to $(2, 2)$, and once at $(2, 2)$ back to $(0, 0)$.'

In their project work, we ask the students to model the above scenario using Uppaal and to check whether Wooldridge's aforementioned claim is correct. Having realised that there is actually *no* deterministic, memoryless strategy for the vacuum cleaner that extends Wooldridge's four rules, the students then find out that one has to remove all but the first rule in order to find a suitable strategy for the robotic agent that meets the stated criteria. Next, we ask the students to consider a number of timing-based scenarios, asking them to find the fastest strategy for the robot to vacuum clean the grid. Having examined a number of specific settings of the time f it takes the robot to move forward and the time r that the robot needs to rotate, and after having mapped all the possible combinations of forward moves and rotations in a strategy, the students quickly realise that a shortest strategy is also a fastest one unless $2r < f$ and that, in that case, a strategy with 16 rotations and 10 forward moves is faster than a shortest one, which involves 12 turns and 12 forward moves.

In order to entice our students to develop scalable models and to reflect on the qualities of their models as a whole, we also ask them to consider the vacuum-cleaning scenario in a 4-by-4 grid and to assess their models vis-a-vis the seven criteria listed by Vaandrager in [17, Sect. 1.10].

Each student group delivers a project report describing their work on the vacuum-cleaning project, together with all their Uppaal models and query files. Again, we devote the start of the course session on the last day of the second week of the course to a 'conference-like session', during which each group of students delivers a short talk presenting their work on the project.

4.3   Week Three: Second Project and Final Exam

The last week of the course is devoted to the second group project and ends with a final, oral exam contributing 30% of the final grade.

The second project we have used most often over the years is based on the solitaire game Rush Hour, which is today produced by Thinkfun. Students are expected to model the game using Uppaal and to use the tool to solve the puzzle for a variety of starting configurations. (See http://www.icetcs.ru.is/remo-course/project2.html for the latest version of the project description.) Apart from being fun, Rush Hour is an example of a challenging solitaire game, which lends itself to a number of variations that can be used to exercise the students' modelling abilities.

First of all, Rush Hour is hard, for humans and machines alike! Indeed, its generalised version is PSPACE-complete [10] and, as recently shown in [7], this hardness result holds true even with only 1×1 cars and fixed blocks. Moreover, the hardest solvable configurations for the game's six-by-six board found by Collette, Raskin and Servais in [8] are fiendishly difficult for humans, involving as many as 93 moves to solve optimally[10]. Second, the game can be modelled in

---

[10] See http://www.icetcs.ru.is/remo-course/RushHourHardestConfigurations1.pdf for a list of the six hardest game configurations.

a number of natural ways and we encourage our students to explore at least two modelling approaches, comparing the ease with which they can be analysed using the Uppaal verifier. We think that this experience is particularly instructive for students, since it helps them realise that apparently innocuous modelling choices can have a huge effect on processing time during their verification and makes them realise the exponential growth in the number of possible interleavings in the executions of concurrent systems in a playful setting. To make state-explosion come to the fore even more, we ask students to model the Rush Hour game in which a step in the game allows one to move a vehicle more than one position at a time.

In order to make students evaluate the extensibility and reusability of their models, we ask them to model the version of Rush Hour with walls and use the Uppaal verifier to solve the hardest such puzzles given at https://www.michaelfogleman.com/rush/.

Finally, we use the game to introduce students to issues related to mutual exclusion in concurrent programming, some of which they will meet in their future courses on Operating Systems and Concurrent Programming. Concretely, we ask them to assume that a two-handed player can move two vehicles (one position) simultaneously. The player is left-handed and takes 3 s to move a vehicle one position with the left hand, but 5 s to move a vehicle one position with the right hand. The keenest students are expected to model this scenario using Uppaal and to find the fastest way of solving the puzzle from some of the starting configurations we provide. This task is optional and open ended; we stress to the students that we are mainly interested in seeing how they approach the modelling task, and in the type of possible pitfalls they identify. Having said so, most student groups over the years have attempted to solve this problem and we even had one group of students that considered a scenario in which the player is an octopus!

The last week of the course ends with a final, oral exam. Even though the course is very much project based and students do report to us when some group members are not contributing to the project work, we still need to check whether some students are free riding. To this end, we examine one student group at the time as follows. The exam session starts with a presentation in which the group being examined presents its work on the second project. We then ask questions to the group based on their work during the three weeks. Initially, questions are addressed to the group as a whole and we give everyone who wants to answer the chance to do so. If we realise that some students do not attempt to answer any of our questions, we direct some questions specifically to them to gauge how much they know about the work their group has done during the course. Each student receives an individual grade for the final exam, whereas work on the projects is graded at the group level.

5 The Two Pandemic Editions

The REMO course was largely held in a 'Renaissance-workshop style'[11] in the period 2013–2019. All student groups worked independently in the university building in a suitably-sized lecture room and in the surrounding areas, as needed. As mentioned above, our role is largely a supporting one. We are there to assist students in their independent learning and work. In those editions of the course, we were available to answer questions from each group during their supervised working sessions in the mornings, and regularly visited each group at that time to check how their work progressed and whether they had any specific issues that were stifling their progress. Whenever problems of general interest arose, we would call all groups for a brief plenary presentation addressing those problems and how they might be tackled using Uppaal, as well as providing pointers to the literature. Student course evaluations over the years indicate that students like this workshop-like setting, and feel empowered by the trust we show in them and the level of independence we expect in their work. To wit, we limit ourselves to mentioning a comment we received from one student, who later asked for a letter of references from one of us:

> I enjoyed taking your course and loved how the barrier between teacher and student was broken during the course. (I have been and will continue to recommend the course to any student at Reykjavik University that I know).

During the pandemic, we had to hold the 2020 and 2021 editions of the course fully online, which were followed by 19 and 24 students, respectively. However, despite the online setting, we strove to maintain the workshop-style structure of the course and to uphold its key pedagogical tenets mentioned in Sect. 2. To this end, we ran the course trying to mimic the in-person sessions in an online setting as faithfully as possible. The course was held on Jitsi and we were available to answer questions from the student groups at the same Jitsi link throughout the course from 8:30 till 17:00, taking shifts with one of our PhD students. We also had a Discord channel for the course and most student groups used Discord to coordinate their project work.

We started the day each morning with a scrum meeting, during which each group would report on the progress they made and on what they planned to do during the day[12]. We alternated between public scrum meetings, in which all groups were virtually present at the same time, and group-specific scrum

---

[11] By this, we mean that our course sessions tried to replicate the creative atmosphere of the busy workshops in which many of the great Renaissance works of art were produced. Those workshops were run by experienced artists, who trained young ones in their crafts and also learnt from the new ideas produced by their trainees.

[12] We also held impromptu scrum sessions at the start of each day in all the in-person editions of the course. However, due to the lack of face-to-face contact with the students and in order to help students structure their remote working day, we felt that we needed to be much more systematic in doing so during the two pandemic editions.

meetings. All conference sessions and the final exams were held online, following the same structure we adopted for the in-presence editions of the course.

Based on the feedback we received from the students who took the pandemic editions of the course, this set-up worked rather well and the students felt that they were making progress independently, while receiving a good level of support from us. They were pleased with the willingness we showed to answer their questions and with our short response time. The following excerpt from a particularly eloquent course evaluation we received from a second-year student in Software Engineering is representative of the students' opinions:

'The course was conducted entirely online, using various platforms to interact with the professors and other students. It was interesting inviting people you just met into your room, virtually, and presenting your work. It takes some time getting used to, but if this past year has taught us anything, it's that working from home is very much possible. Inevitably, some people had minor technical issues but overall, I believe that this aspect of the course, as well as all other aspects, was a success and think that we all learned something new in these three weeks. Even the professor took on a new communication tool he had never used before which was popular among the students. It's all about patience and willingness to adapt to a new environment'.

Based on that circumstantial evidence, it seems that we did manage to meet our goals to hold an engaging, student-driven course that makes them responsible for their own learning. However, the price we had to pay to do so was to devote more time and energy to the course than we typically do in an in-person setting. Even though it may seem counter-intuitive, empowering students in an online setting required more involvement from us during the afternoon course sessions than it did in an in-person setting, where students work in the university building with little or no assistance from us. Based on discussions with our colleagues both at our institution and elsewhere, as well as on the literature on teaching during a pandemic we have read over the last year (see, for instance, [9,11,12]), it is fair to say that we are not alone in feeling that way. To wit, even though such figures have to be taken with a grain of salt, it is estimated that teaching online can roughly triple the preparation time for a one-hour lecture [12]. Moreover, according to a survey mentioned in [9], university lecturers overwhelmingly thought that the pandemic has made their job more difficult than it was before, with 87% of the respondents either strongly agreeing or agreeing. Some of the possible reasons for the increased workload on teachers during the pandemic are mentioned in [11, Sect. 4.3.2]; these include the need to record lectures in advance and edit them thereafter to clarify and/or correct some of the material, the interaction with learning management systems that sometimes needed to be adjusted to support teaching taking place fully online, the use of more time for grading assignments since students ended up working alone on tasks that could be carried out in groups, and answering the same questions repeatedly and on a variety of tools used to communicate with the students taking a course.

6   Evaluation and Conclusions

The REMO course has run at Reykjavik University in the three-week part of the spring semester every year since 2013. It has been followed by roughly 300 students overall and has consistently received high grades from the students in their course evaluations. To wit, the average evaluation of the students' satisfaction with the course lecturers since 2013 is 4.9 out of 5. The average score of the course in the period 2013–2021 in the other evaluation criteria adopted by Reykjavik University is as follows:

- Course satisfaction: 4.79 out of 5.
- Assessment of the outcome of the course: 4.61 out of 5. (This metric refers, for example, to factors such as overall understanding of course content, increased interest in the academic field and study materials, and personal learning success).
- Course organisation: 4.84 out of 5. (This metric refers, for example, to factors such as the teaching plan, learning objectives, course assessment, planning of classes, and organisation of assignments).
- Evaluation of how the course assignments helped in increasing student understanding of the study material: 4.79 out of 5.

(Of course, pleasing as they may be, the results of these student course evaluations should be considered merely as a measure of student satisfaction with the course. As indicated in, for instance, the meta-analysis of faculty's 'teaching effectiveness' reported in [16], that metric is not related to student learning).

Most importantly, we believe that the course has made the students taking it aware of the existence of mature and usable modelling and verification technology developed by the research community working on formal methods over many years. Those students now have a basic understanding of what formal methods can (and cannot) do, know that they are routinely employed within the high-technology companies whose products they use daily, and have repeatedly told us that they will consider using model checkers in their future studies and careers as appropriate. Moreover, they serve as ambassadors for the course, and hopefully also for formal methods, within the rest of the student population and in their future workplaces. By way of example, we limit ourselves to citing two comments we received from students in their course evaluations:

'Very interesting course, introducing techniques and tools that I think it is likely that I will use in the future. I chose the course blindly, having no idea of what formal methods and model checking are. I was positively surprised by what I learnt. I will recommend this course to my colleagues. (A second-year student in Computer Science)'

'At the start of the course, I had absolutely no knowledge on the subject.... The main takeaway was that using human ingenuity only takes you so far when optimizing complex systems and it can be difficult to prove that you have the most efficient design without using the proper tools and methods we learned about in the course.... The work I did in this course is some of my proudest work during my studies. (A second-year student in Software Engineering)'

We think that 'spreading the formal-methods gospel' is important at a department like ours, since far too many of our students graduate without taking a single course that exposes them to formal modelling and verification. One of the roles of the REMO course is to try and change this state of affairs, but it is fair to say that we still have work to do, despite our efforts since 2013. For instance, the number of students taking the course is still fairly small (ranging from 18 in 2013 to 36 in 2016) and increasing it will be non-trivial in the light of some compulsory three-week courses that students in Computer Science and Software Engineering take in the spring semester. So far, first-year students for which REMO is compulsory account for 46% of all the students who have taken the course, 19% have been second-year students and 29% have been third-year or exchange students. Since the Discrete Mathematics and Computer Science degree course is a niche study programme, it is unlikely that the percentage of first-year students taking the course will increase noticeably. Therefore, we will target third-year students, even though some of them might be finalising their final bachelor projects in the spring semester.

We might also consider turning REMO into a typical twelve-week course, while maintaining its philosophy. We believe that this is doable by identifying a suitable number of modelling and verification problems that student groups can solve in two-to-four-hour, supervised study sessions. However, doing so would dilute the students' study experience somewhat, the course would lose its high-intensity flavour and we would have to compete for the students' brain cycles with the other courses running in parallel with that version of the REMO course.

Moreover, since the course has run since 2013, we should carry out a data-driven analysis of how well it achieves its learning outcomes and high-level goals. For instance, it would be interesting to collect and analyse data to try and understand whether students taking our course find that the introduction to formal methods they received has had an impact on their practice. Furthermore, we should collect student feedback on whether our 'warm-up week' does play a role in creating a nurturing and positive learning environment at the start of the course.

In conclusion, based on our experience, one can introduce formal methods to early-career bachelor students in high-intensity-training mode in as little as three weeks. Our REMO course provides a possible blueprint for doing so, while empowering students to learn and work independently, and challenging them to step out of their intellectual comfort zone and to take charge of their own development. We hope that our experience can be useful to others.

Acknowledgements. We are grateful to the anonymous reviewers and the PC chairs, who offered some detailed and insightful comments and suggestions that led to several improvements in our original manuscript. We thank Elli Anastasiadi for her excellent work as student mentor during the last three editions of the REMO course. In particular, Elli's assistance has been invaluable during the two pandemic installments in the spring of 2020 and 2021. We are most grateful to Kim G. Larsen for the Uppaal-related course material he has generously shared with us over the years and to Marius Mikucionis for the sterling support he has offered our students and us on Uppaal-related matters. Catia Trubiani provided useful feedback on a draft of this paper. Our colleagues Nidia Guadalupe López Flores and María Óskarsdóttir pointed out some interesting papers on teaching during the pandemic. Last, but not least, we thank Hildur Daví sdóttir for eloquently sharing her opinions on the 2021 edition of the REMO course with us. Any remaining infelicity is solely our responsibility.

References

1. Aceto, L., Ingólfsdóttir, A., Larsen, K.G., Srba, J.: Reactive Systems: Modelling, Specification and Verification. Cambridge University Press, Cambridge (2007)
2. Aceto, L., Ingolfsdottir, A., Larsen, K.G., Srba, J.: Teaching concurrency: theory in practice. In: Gibbons, J., Oliveira, J.N. (eds.) TFM 2009. LNCS, vol. 5846, pp. 158–175. Springer, Heidelberg (2009). https://doi.org/10.1007/978-3-642-04912-5_11
3. Alur, R., Dill, D.L.: A theory of timed automata. Theor. Comput. Sci. **126**(2), 183–235 (1994). https://doi.org/10.1016/0304-3975(94)90010-8
4. Arnold, A., Bégay, D., Crubillé, P.: Construction and Analysis of Transition Systems with MEC. AMAST Series in Computing, vol. 3. World Scientific (1994). https://doi.org/10.1142/2505
5. Behrmann, G., David, A., Larsen, K.G.: A tutorial on Uppaal. In: Bernardo, M., Corradini, F. (eds.) SFM-RT 2004. LNCS, vol. 3185, pp. 200–236. Springer, Heidelberg (2004). https://doi.org/10.1007/978-3-540-30080-9_7
6. Behrmann, G., David, A., Larsen, K.G., Pettersson, P., Yi, W.: Developing UPPAAL over 15 years. Softw. Pract. Exp. **41**(2), 133–142 (2011). https://doi.org/10.1002/spe.1006
7. Brunner, J., et al.: 1 × 1 rush hour with fixed blocks is PSPACE-complete. In: Farach-Colton, M., Prencipe, G., Uehara, R. (eds.) 10th International Conference on Fun with Algorithms, FUN 2021, 30 May–1 June 2021, Favignana Island, Sicily, Italy. LIPIcs, vol. 157, pp. 7:1–7:14. Schloss Dagstuhl - Leibniz-Zentrum für Informatik (2021). https://doi.org/10.4230/LIPIcs.FUN.2021.7
8. Collette, S., Raskin, J.-F., Servais, F.: On the symbolic computation of the hardest configurations of the RUSH HOUR game. In: van den Herik, H.J., Ciancarini, P., Donkers, H.H.L.M.J. (eds.) CG 2006. LNCS, vol. 4630, pp. 220–233. Springer, Heidelberg (2007). https://doi.org/10.1007/978-3-540-75538-8_20
9. Flaherty, C.: Faculty pandemic stress is now chronic. Inside Higher Ed, November 2020. https://www.insidehighered.com/news/2020/11/19/faculty-pandemic-stress-now-chronic
10. Flake, G.W., Baum, E.B.: Rush hour is PSPACE-complete, or "why you should generously tip parking lot attendants". Theor. Comput. Sci. **270**(1–2), 895–911 (2002). https://doi.org/10.1016/S0304-3975(01)00173-6
11. Flores, N.G.L., Islind, A.S., Óskarsdóttir, M.: Effects of the COVID-19 pandemic on learning and teaching: a case study from higher education. CoRR abs/2105.01432 (2021). https://arxiv.org/abs/2105.01432
12. Gewin, V.: Pandemic burnout is rampant in academia. Nature **591**, 489–491 (2021). https://doi.org/10.1038/d41586-021-00663-2

13. Hamberg, R., Vaandrager, F.W.: Using model checkers in an introductory course on operating systems. ACM SIGOPS Oper. Syst. Rev. **42**(6), 101–111 (2008). https://doi.org/10.1145/1453775.1453793
14. Hurkens, C.: Spreading gossip efficiently. Nieuw Arch. voor Wiskunde **5/1**(2), 208–210 (2000). http://www.nieuwarchief.nl/serie5/pdf/naw5-2000-01-2-208.pdf
15. Rozier, K.Y.: On teaching applied formal methods in aerospace engineering. In: Dongol, B., Petre, L., Smith, G. (eds.) FMTea 2019. LNCS, vol. 11758, pp. 111–131. Springer, Cham (2019). https://doi.org/10.1007/978-3-030-32441-4_8
16. Uttl, B., White, C.A., Gonzalez, D.W.: Meta-analysis of faculty's teaching effectiveness: student evaluation of teaching ratings and student learning are not related. Stud. Educ. Eval. **54**, 22–42 (2017). https://doi.org/10.1016/j.stueduc.2016.08.007
17. Vaandrager, F.W.: A First Introduction to Uppaal. https://www.mbsd.cs.ru.nl/publications/papers/fvaan/handbookuppaal/. to appear in Quasimodo Handbook, J. Tretmans editor
18. Vaandrager, F.W., Verbeek, F.: Recreational formal methods: designing vacuum cleaning trajectories. Bull. EATCS **113** (2014). http://eatcs.org/beatcs/index.php/beatcs/article/view/269
19. Wooldridge, M.J.: An Introduction to MultiAgent Systems, 2nd edn. Wiley, Hoboken (2009)

Online Teaching of Verification of C Programs in Applied Computer Science

Matthias Güdemann[✉][iD]

University of Applied Sciences (UAS) Munich, Munich, Germany
`matthias.guedemann@hm.edu`

Abstract. This is a report on teaching formal methods in the form of program verification for Master students in an applied computer science setting. The course was taught fully online, using recorded videos, synchronous sessions, the learning management system Moodle (https://moodle.org/), a distributed version control system and mostly biweekly graded practical assignments.

The first objective was to use the C language. It is a very relevant language in the sectors where verification is used in industry. The students already know the language, it also has interesting properties which can make verification challenging and shows the importance of edge cases in verification. The second objective was to teach the use of mature, industrial-strength tools in order to make the skills transferable to the later work situation of the students. This required tools that are actually used in industry to analyze C programs. The third objective was to introduce different verification approaches and to show the strengths and potential limitations of each. The selected approaches were deductive verification, abstract interpretation and model checking.

To achieve these goals, Frama-C with its WP and EVA plugin, the model checker CBMC and the Z3 SMT solver were selected. Because of the applied setting it was desired to use examples which did not require the use of interactive theorem proving for deductive verification.

1 Introduction

Teaching formal methods and program verification is an important part of computer science education. Just writing specifications of programs is often hard for the students and having to do so is a very valuable experience in its own. Being able to prove properties of programs gets more and more widespread in many domains, in particular as program security gets ever more important.

At the same time, formal methods are often considered to be a theoretic or academic topic without clear application in practice. In particular in an applied computer science setting where the focus is less on research and theoretical foundations and more on applicable topics. Often it is also functional programming languages and dependent types which are used in program verification. This has the advantage of having the Curry-Howard isomorphism as a clear correspondence between programs and proofs. The downside is that while functional programming aspects are used more and more in modern programming languages,

© Springer Nature Switzerland AG 2021
J. F. Ferreira et al. (Eds.): FMTea 2021, LNCS 13122, pp. 18–34, 2021.
https://doi.org/10.1007/978-3-030-91550-6_2

programming purely in functional languages is still limited to very few niches. Therefore, it is unlikely that many of the students will do this later in their jobs.

Choosing C as a language for program verification has some extra challenges but also benefits. C has a lot of rather special and difficult aspects. At the same time it is widely used in domains where program verification is mandatory. Also, there exist different industrial strength verification tools for C which allows for hands on experiments in verification on real programs.

The rest of the paper is structured as follows: Sect. 2 gives some background on the university, C verification and the lecturer. Section 3 introduces the verification tools that were selected for the course. Section 4 gives some detail on the online teaching setting and Sect. 5 gives an overview of the exercise assignments the students had to complete. Section 6 reports on the challenges the students faced and the evaluation of the course by the students. Section 7 concludes the paper with some outlook on the changes for the next iteration of the course.

2   Background

2.1   University of Applied Sciences

The German University of Applied Science (UAS) is traditionally a type of university with a principal focus on teaching applied topics. For the professors at UAS it is a requirement to have worked at least 3 years outside academia and most have at least 5 years of industrial experience. Teaching is generally focused on current topics and with applicability in industry in mind.

In recent years the image of UAS is changing and the focus changes in the direction of research, in particular applied research, i.e., topics which show promise of commercial use in a short time-period. This shift to research has also changed the topics that are taught in computer science curricula. In Munich this has led to the introduction of formal methods teaching in the form of program verification and model-checking. Such a focus is still rather uncommon at UAS but the experience shows that the students do see the merits of formal methods if taught in a way that shows real applicability.

2.2   C Program Verification

Choosing C as the target language for program verification was a compromise between the complexity of the properties to show and the applicability of the topic in a later industrial setting. C is still a widely used language in embedded systems and safety critical domains which is an important domain for program verification.

The complexity of the properties is limited due to C not being designed with verification in mind. On the contrary, C has aspects like undefined or implementation defined behavior which makes verification tricky. At the same time this offers interesting topics for discussion in the classroom when implicit assumptions the students have turn out to be false.

Using C for verification can be applicable in an industrial context because C is still widely used. This is because compilers exist for almost every architecture and also because it allows very fine-grained control which can be essential in embedded systems or low level development like operating systems. This important position of C also means that there exist industrial grade verification tools. Learning these tools is also a potential source of readily applicable knowledge for the students.

3   Verification Approaches and Tools

The choice of verification approaches and of tools is closely connected. It was important to select modern tools that are capable of reading real C programs and not just subsets of C or abstracted languages which lack some of the really challenging aspects. We chose verification approaches, deductive verification, abstract interpretation and software bounded model-checking. This was done to show their respective strengths and weaknesses of these approaches.

Because of the curriculum of the UAS there was no real background in mathematical logic, therefore it was important have more or less full automation support even for deductive verification. Introducing interactive theorem proving would have taken too much time for the course.

The Frama-C framework [3, 10] offers support for expressing properties in the Ansi C specification language (ACSL) [2]. This provides a good integration into C programs as annotations of functions or definitions of logic functions. At the time of the course the current version of Frama-C was 21.0. C programs with ACSL annotations can be compiled and executed just as normal C programs without these annotations. All ACSL annotations are expressed in specially formatted comments.

3.1   Deductive Verification

For deductive verification the Frama-C/WP plugin was chosen. It uses ACSL specifications of functions and uses preconditions and loop annotations to create proof obligations to prove the properties. The proof obligations can be discharged by different external tools. Frama-C uses the Why3 platform [9] which supports different SMT solvers, first order logic theorem provers or external interactive theorem provers. There is an extensive work on formalization and verification of standard algorithms in C available [6].

The following code illustrates an ACSL annotation in the form of a two part loop invariant and a loop variant.

```
/*@
  loop invariant 0 <= i <= n;
  loop assigns i;
  loop variant n - i;
*/
for(int i = 0; i < n; i++) { ... }
```

The following code illustrates an ACSL annotation of a function contract. It specifies that the postcondition of the function is that b points to the original pointer of a and that a points to the original pointer of b, i.e. the values at the pointers have been exchanged when the function returns. The original pointer is marked as *old*, one can specify own labels in addition to standard labels that are available.

```
/*@
  ensures *b == \old(*a);
  ensures *a == \old(*b);
*/
void swap (int *a, int *b) {
  int tmp = *a;
  *a = *b;
  *b = tmp;
}
```

3.2 Abstract Interpretation

For abstract interpretation the Frama-C extended value analysis (EVA) [4] plugin was chosen. It allows for fully automatic verification of ACSL specifications using abstract interpretation and related techniques. It uses internal abstract domains and different options that control the analysis, e.g., the number of internal states to analyze.

The advantage of this kind of analysis is that if no error is reported, then no runtime error is possible in the program. If an error is reported, this means that in the abstraction a runtime error can occur, so then there *may* be a problem in the concrete program.

```
int f(int a) {
  int x, y, sum, result;
  if(a == 0) {
    x = 0; y = 5;
  } else {
    x = 5; y = 0;
  }
  sum = x + y;
  result = 10 / sum;
  return result;
}
```

Given the above code the plugin manages to prove the absence of a runtime error in the form of division by zero. This is shown in the below output of Frama-C. While in the resulting value sets the variables x and y can have either value 0 or 5, the value of $sum = x + y$ cannot have the value 0 because x and y cannot have value 0 at the same time.

```
[eva] done for function f
[eva] ====== VALUES COMPUTED ======
[eva:final-states] Values at end of function f:
  x in {0; 5}
  y in {0; 5}
  sum in {5}
  result in {2}
[eva:summary] ====== ANALYSIS SUMMARY ======
  ------------------------------------------------------------------
  1 function analyzed (out of 1): 100% coverage.
  In this function, 8 statements reached (out of 8): 100% coverage.
  ------------------------------------------------------------------
  No errors or warnings raised during the analysis.
  ------------------------------------------------------------------
  0 alarms generated by the analysis.
  ------------------------------------------------------------------
  No logical properties have been reached by the analysis.
  ------------------------------------------------------------------
```

3.3   Software Bounded Model-Checking

For software bounded model-checking the CBMC [11] tool was chosen. It allows for specification of assertions directly in C code and uses bit-precise model-checking using SAT and SMT solving to verify or disprove the properties. From a teaching point of view, the possibility to get counterexamples for false properties is a very interesting feature of model-checking.

In addition to CBMC, pure SMTLIB2 SMT solving in the form of Z3 [8] and CVC4 [1] was used to illustrate the formalization of lemmas in the form of satisfiability of constraint problems. Frama-C/WP also uses SMT solvers for deductive verification, but the proof obligations and encoding of these problems were not detailed in the lecture.

CBMC allows for textual generation of verification conditions in different formats. This is a useful feature to illustrate how the C programs are transformed into single static assignment (SSA) and then translated into constraint problems for SMT solving.

In the following code the assertion specifies that the sum of parameter x and y cannot be zero.

```
int f(int n, int x, int y) {
  int divisor = 0x12345678 - x + (y << 1);
  assert (divisor != 0);
  return n / divisor;
}
```

CBMC translates this into the following constraint problem. This is direct output of CBMC, slightly shortened to reduce it to the essential part.

```
{-22} f::n!0@1#1 = nondet_symbol
{-23} f::x!0@1#1 = nondet_symbol
{-24} f::y!0@1#1 = nondet_symbol
{-25} f::1::divisor!0@1#2
      = 305419896 + shl (f::y!0@1#1, 1) + -f::x!0@1#1
|----------------------
{1} ¬(f::1::divisor!0@1#2 = 0)
```

The first 3 lines of the constraint problem state that the three parameters of the function f are equivalent to a nondeterministic value, i.e., the constraints 22, 23 and 24 are always fulfilled. The constraint 25 then defines that the local variable *divisor* of the function f is equal to the right side which corresponds to $305419896 + (y \ll 1) - x$. These constraints describe the program in single static assignment (SSA) in the form of equivalences that relate the variables and parameters.

From these constraints CBMC then tries to deduce the property in the form of the assertion $\neg(divisor = 0)$. This is done by negating the property, i.e., adding the additional constraint $divisor = 0$. An SMT solver then checks satisfiability of the conjunction of the constraints and the negated property. If this is satisfiable then there exist parameter n, x and y such that $divisor = 0$. A satisfying assignment comprises a counterexample to the property.

For this program and property it is of course possible to choose the function parameters in such a way that a division by zero is possible. The following shows the counterexample as generated by CBMC. This means that with x equal to -1841559528 and y equal to 1073993936 the calculated value of *divisor* is 0, the value of n is not important. The possibility to get counterexamples is very useful: "It is impossible to overestimate the importance of the counterexample feature" [7].

```
State 33 file div.c function __CPROVER__start line 5 thread 0
----------------------------------------------------------
  n=0 (00000000 00000000 00000000 00000000)

State 34 file div.c function __CPROVER__start line 5 thread 0
----------------------------------------------------------
  x=-1841559528 (10010010 00111100 00001000 00011000)

State 35 file div.c function __CPROVER__start line 5 thread 0
----------------------------------------------------------
  y=1073993936 (01000000 00000011 11011000 11010000)

State 36 file div.c function f line 6 thread 0
----------------------------------------------------------
  divisor=0 (00000000 00000000 00000000 00000000)
```

```
Violated property:
  file div.c function f line 7 thread 0
  assertion divisor != 0
  divisor != 0
```

4 Online Teaching

Due to the restrictions because of the COVID-19 pandemic, the course was taught fully online. In addition, because of the restrictions concerning in-person exams, grading was done on practical exercises. Each exercise was for around 2 weeks and could be completed in teams of two students or alone.

Online teaching worked quite well. The exercises were organized via github classroom[1] which allows for easy creation of repositories from templates for the students.

The course was held in the following way: each week there was an asynchronous part where new material was distributed as recorded videos and slides. At the normal lecture date there was a synchronous session where the material was presented in more detail. In the synchronous part the students were also asked to complete several small multiple-choice quizzes per session. Most sessions also included live-demos of the relevant aspects of the currently used tools.

We also employed Rocket.Chat[2] which proved to be very helpful to exchange code snippets to discuss problems or questions for the practical exercises. It also integrates Jitsi[3] to support video calls and live screen sharing.

To prevent most kinds of compatibility problems we decided to use a standardized virtual machine as the software platform. The VM was based on a standard Ubuntu Linux with the different tools preinstalled. Frama-C can easily be installed via the *opam* package manager for OCaml. CBMC and Z3 are directly available as packages in Ubuntu.

5 Exercise Selection

Due to the fact that a basic course in mathematical logic was not compulsory in the students' curriculum it was necessary to start with basics of specification using first order logic. From these foundations we then switched to formal specification and Hoare logic.

For each exercise we give a short paragraph on the preceding preparation lectures, the task itself and the goal of the exercise. Tools like Frama-C, SMT solvers and CBMC were presented in live-demo sessions in the synchronous sessions.

---

[1] https://classroom.github.com.
[2] https://rocket.chat/.
[3] https://jitsi.org/.

5.1   Exercise 1—Informal Specification

graded no/**time** 1 week

Preparation. In the lecture before this exercise the students got an introduction to propositional logic.

Task. The first exercise was to clone a repository which contained a single C file and to analyze informally what the function in the C file would do. The students were asked to compile the file, execute it and to validate their guess what the function f computes.

```
int f(int n) {
  int s = 0;
  int i = 1;
  while (i <= n) {
      s = s + i;
      i++;
  }
  return s;
}
```

The next step was to write down a specification of what the function computes. This specification was intended to be informal and it also was the first time the students had to write a specification on their own.

Finally, the students were asked to think about edge cases for which the function might not fulfill the specification.

Goal. The intention of this exercise was to familiarize the students with C programs, to get an idea about the difficulties to express precisely what a function is intended to do and also with the fact that machine integers do not always behave like unbounded integers.

5.2   Exercise 2—First Order Logic

graded no/**time** 1 week

Preparation. In the lecture before this exercise the students got an introduction to first order logic with many different examples of formalized properties.

Task. The next exercise was to express the specification of exercise 1 as a first order logic formula. Still, in free-form, not yet in a standardized way like ACSL.

Goal. The intent here was to familiarize the students with the challenge to use logic to correctly specify a property.

5.3   Exercise 3—Hoare Logic

graded yes/**time** 1 week

Preparation. In the lecture before this exercise the students got an introduction to different approaches to program testing and coverage criteria, proof trees and Hoare logic. For Hoare logic reasoning a simple imperative language was introduced to explain the separate rules for the different language constructs.

Task. This exercise was the first graded exercise in the course. It included simple properties which had to be proven using Hoare logic and manually writing down the proof tree of the Hoare rule applications.

This included the calculation of the weakest precondition of the following programs.

```
// which precondition is required for postcondition y > 1?
y := x + 1;

// which precondition is required for postcondition z > 0?
y := x + 1;
x := y + 1;

// which precondition is required for postcondition z > 0?
z := x * y;
```

The next part was to specify a loop invariant such that with the precondition $n \geq 0$ the postcondition $acc = 2 * n$ holds. It was also asked to give the loop variant which guarantees termination.

```
acc := 0;
i := 0;
while (i < n)
  acc := acc + 2;
  i := i + 1;
```

The last part then asked to generalize the *while* rule to a rule for *for* loops, i.e. to specify how a Hoare-style rule for *for* loops would have to look like in order to prove correctness of the Hoare triple.

Goal. The intent of this exercise was to familiarize the students with Hoare logic reasoning which is at the base of deductive verification. Loop invariants (and variants) generally have to be specified manually. Understanding how the Hoare logic rules for loops work is a very important concept in verification of imperative programs.

5.4   Exercise 4—Deductive Verification Using Frama-C

graded yes/**time** 1 week

Preparation. In the lecture before this exercise the students got an introduction to the ACSL specification language for C and to Frama-C. This consisted mainly of the ACSL operators for first order logic and the specific keywords to express function preconditions, properties, assertions and loop invariants.

For Frama-C this included running the command line version of the tool on an annotated C file and the interpretation of the resulting output messages.

Task. The next exercise was based directly on exercise 3. For this exercise, the programs were given as C programs, the pre- and postconditions and the loop invariants had to be expressed as ACSL annotation before and after each statement, corresponding to the Hoare triple. The goal was to prove the postconditions from the specifications and loop invariants using Frama-C/WP in a fully automatic way. The exercise also included the formalization of a lemma for multiplication and to check which of the SMT solvers was capable to verify the lemma automatically.

Goal. The intent of this exercise was to familiarize the students with ACSL specifications and with using Frama-C. We limited the use to the command line interface which is more than adequate for tasks like these. The exercise did not yet use fixed-width machine integers, any runtime warnings were to be ignored.

5.5   Exercise 5—Arrays

graded yes/**time** 2 weeks

Preparation. In the lectures before the students got a reminder on peculiarities of the C language, in particular pointers, as well as an introduction on control flow graphs (CFG) as program representations.

Task. The next exercise dealt with more complex specifications. There were three parts. The first part was to specify the return value of a function that computes the minimum of two integers.

```
int min(int x, int y) {
  int z = x < y ? x : y;
  return z;
}
```

The second part was the first task to include arrays. For a given array of integers and its length, the index of the minimal element was to be returned. If no such element exists, then a special value had to be returned.

```
int min_array(int* arr, int len) {
  if (len == 0)
    return -1;

  int min = 0;

  int i;
  for (i = 0; i < len; i++) {
      if(arr[i] < arr[min])
         min = i;
    }
  return min;
}
```

The third part consisted of finding the smallest non-negative value in a sorted array. The specification here included specifying that the values in the array are sorted in a non-decreasing order.

```
int min_pos_array(int* arr, int len) {
  if (len == 0)
    return -1;

  for (int i = 0; i < len; i ++) {
      if (arr[i] >= 0)
         return i;
    }
  return -1;
}
```

Goal. The intent of this exercise was to familiarize the students with function contracts in ACSL in addition to the statement annotations. Already the specification of minimum is non-trivial, several solutions only specified that the return value should be less than or equal to both input parameters.

For the sorted array, several of the students specified a pairwise predicate, comparing only direct successor elements. This led to problems with the automatic provers. The SMT solvers required a global specification of a sorted array in order to provide fully automated proofs. This illustrated the difference between a specification which is good for verification and a specification which would be easy to translate into an efficient implementation.

5.6   Exercise 6—Runtime Errors

graded yes/**time** 2 weeks

Preparation. In the lecture before this exercise the students got an introduction into the possible runtime errors of C programs. This also included the different warnings that Frama-C/WP produces to prevent runtime errors from appearing. This includes mainly integer overflow/underflow, pointer validity and aliasing.

Task. The next exercise was split into two parts. The first part was to add preconditions to most of the former exercises in such a way that any runtime errors were excluded. Frama-C/WP provides an option to generate proof-obligations for showing the absence of runtime errors.

The second part was the implementation and specification of a variant of the famous *fizz-buzz* program. In this form it incorporated 3 arrays of same length. At each index divisible by 3 and 5 the first of the arrays should have a value 1 and the two others a value 0. At each index divisible by 3 the second should have value 1 and at each index divisible by 5 only the third array should hold the value 1.

Goal. The intent of this exercise was to familiarize the students with all different kinds of possible runtime errors in a language like C. This does not only include potential integer overflow or illegal memory access, but also aliasing in form of overlapping arrays.

A fully complete and correct specification of the fizz-buzz function proved to be tricky. The main implementation variant was first to zero all arrays and then fill the arrays with values 1 where appropriate. Unfortunately this solution requires a more complex loop invariant than using a single loop and filling each array at each index with the correct value 0 or 1. It also illustrated well that specifying *what* a program does exactly can be more difficult than implementing this functionality.

5.7 Exercise 7—Abstract Interpretation

graded yes/**time** 1 week

Preparation. In the lectures before the students got an introduction to abstract interpretation. This includes a simple sign domain as example and an overview of different properties that can be verified by abstract interpretation.

Task. The next exercise dealt with abstract interpretation (AI). Frama-C provides the EVA plugin which does a form of AI. Unfortunately, from a didactic perspective, this plugin is quite advanced and it is not possible to reduce the domains to simple ones like the sign domain only. While it is possible to deactivate the normal C-domain, this is discouraged by the authors of Frama-C because it is unlikely to work as expected. There seems to be a gap in the set of analysis tools for C which are based on abstract interpretation which are well adapted for teaching.

Therefore, the exercise itself was divided in a theoretical and a practical part. In the theoretical part, the students were asked to define an abstract domain using first unbounded integer intervals and to define abstract addition and multiplication for this domain. Then the domain changed to fixed bit-width integer intervals with the same task and to note the differences.

The practical part consisted of working through the Frama-C/EVA tutorial [5]. This provides some insight into how such a tool can be used to analyze C code and how to understand the functioning of an unknown program, but it allows for less learning how AI works for real programs.

5.8   Exercise 8—Bounded Model Checking

graded yes/**time** 2 weeks

Preparation. In the lectures before this exercise the students got an introduction to SMT solving and bounded model checking. Specifically the SMTLIB2 format was presented as a standard interchange format for modern SMT solvers.

For bounded model checking, the single static assignment (SSA) form was introduced and it was shown how a program in this form can be expressed in SMTLIB2 using different underlying logics.

Finally, the CBMC tool was presented with the required options like loop unwinding. It was also shown how assumptions can be used to formalize specific properties and how standard coverage properties can be generated based on CFG representations.

Task. The last exercise dealt with bit-precise model-checking of C programs in the form of CBMC and formalizing a problem directly as SMT constraint problem. The first part of the exercise was to analyze the famous *fast inverse square root* program used in a popular 3D-shooter game in 1999[4]. The students were asked to formalize the property that the relative error of this routine was below a threshold for an interval of possible input parameters. The following code[5] is available under a GPL license, the comments have been removed.

```
float Q_rsqrt( float number )
{
        long i;
        float x2, y;
        const float threehalfs = 1.5F;
        x2 = number * 0.5F;
        y  = number;
        i  = * ( long * ) &y;
        i  = 0x5f3759df - ( i >> 1 );
        y  = * ( float * ) &i;
        y  = y * ( threehalfs - ( x2 * y * y ) );
        return y;
}
```

---

[4] https://en.wikipedia.org/wiki/Fast_inverse_square_root.
[5] https://github.com/id-Software/Quake-III-Arena/blob/master/code/game/
q_math.c.

In the second part of the exercise the students were asked to specify an invariant and to verify using CBMC that the following C program computes the absolute value of a given input `float`.

```
float myabs(float v) {
  if(v < 0)
    return -v;
  else
    return v;
}
```

The last part of the exercise was to formalize and prove the following lemma as an SMT problem in `QF_FP` logic, where *float* is the set of 32-bit IEEE 754 floating-point values.

$$\forall x, y \in float : x \times y < 0 \rightarrow (x < 0 \vee y < 0)$$

Goal. The intent of this exercise was to familiarize the students with the way how invariants (in the form of assertions) can be formalized to prove non-trivial properties of C programs. The formalization of a lemma in the form of a satis-fiability problem was intended to familiarize the students with the approach to prove a property by showing that the negation is unsatisfiable.

6   Evaluation

Overall the course worked quite well. There were almost no technical problems, mainly due to the fact that a pre-installed virtual machine was provided. The performance of the VM was more than enough, in particular with hardware-accelerated virtualization.

6.1   Challenges for Students

A big challenge for the students was to understand the reason why deductive proofs did not work. There are mainly two reasons for this: i) the property might not be fulfilled or ii) the pre-conditions or loop invariants are not strong enough to prove the post-condition.

Frama-C lists the proof-obligations which cannot be discharged. The challenge is that these are reported in the form of first order logic which is non-trivial to map back to the original C source code. The main options to alleviate that problem is to use named annotations which allows for more fine-grained report-ing. It is also possible to use the Frama-C GUI which shows the annotations at the source code. Still, in both options the proof obligation is provided encoded in first order logic which is non-trivial to understand.

Overall it is clear that deductive verification is very powerful and at that same time also quite challenging to do. In particular for loop invariants the automated tool support is limited. In the end it is necessary to fully understand the program and also to understand the peculiarities of the C language. Therefore, we consider this more a feature than a problem, verification requires understanding of both the problem and the programming language in order for someone to be able to formalize and solve a problem and then to prove the correctness thereof.

6.2   Results

The overall results of the course were quite good. All students that participated in doing the exercises passed the course. The grades start at 1.0 (best) and go down to 4.0 (worst), 5.0 represents a failure to pass the course. The average grade was 1.52 with the worst grade being 2.3.

The traditional format is to have a written or oral test at the end of the semester. Due to the pandemic this was changed to grade the exercises directly and do a short interview to check whether the students did the work themselves. In these interviews it often became obvious for the students why certain properties would not be proven or what was lacking to have a fully correct specification.

Runtime errors due to overflow seem to be a common knowledge, understanding those did not pose any difficulty to the students. The main challenges were correctly specifying properties and understanding aliasing in C.

For correctness of specifications it might make sense to stress more to check that wrong results are actually not validated. An example would be to show that a specification does not validate an incorrect input to fizz-buzz where the respective entries hold a value of 1 but the others are not necessarily 0 (a rather common error in the specifications).

6.3   Student Evaluation of the Course

At the UAS Munich, every course is evaluated by the participating students. The evaluation is done close to the end of the semester, but before the final test and therefore before the grades are known. Overall 11 of the 17 students in total in the course did respond to the survey. The full results are available in German[6], a summary is shown in Table 1 and Table 2, numbers represent the percentage of the students.

Table 1. Summary of student responses for the course

	Too small	Small	Good	Much	Too much
The amount of learning matter is	0	9.1	81.8	9.1	0
The pace of the course is	0	0	100	0	0
For me the requirements are	0	9.1	81.8	9.1	0
The share of self-learning is	0	0	90.9	9.1	0

---

[6] https://guedemann.org/downloads/Evaluierung_Programmverifikation.pdf.

Table 2. Summary of student responses for their experience

	Very negative	Negative	Neutral	Positive	Very positive
I find the topic is more interesting than I did before	0	0	0	45.5	54.5
I learned a lot in the course	0	0	0	36.4	63.8
I enjoy participating in the course	0	0	9.1	36.4	54.5
I would recommend the course	0	0	9.1	0	90.9
Rating I would give to the course	0	0	0	20	80

Having the standardized virtual machine and software installation allowed for live demos and parallel participation of the students. Exchanging code via Rocket.Chat proved to be very efficient in comparison with screen sharing. Most code used in program verification is rather short, so exchange via text is feasible. Working mostly with command line tools (Frama-C has a GUI but use was mostly text oriented) proved well suited to this mode of online teaching.

This view was shared by the students, one of the free text evaluations said:

"The communication via Rocket.Chat works very well, better than expected. I like the polls during the synchronous lecture as this invites active participation."

7 Conclusion and Outlook

Overall the course and its content was well-received by the students. Using C as language for verification made the course interesting because of the applicability to real world programs. Choosing Frama-C and CBMC as the main tools for the analysis and verification proved to be beneficial as these are mature, industrial strength tools. Using a virtual machine greatly reduced the technical problems with installation and compatibility.

The choice of exercises was for the most part rather conventional and probably more on the easy side. A next iteration should probably include one or two more challenging tasks than this first time. Unfortunately it is not easy to choose good exercise problems for deductive verification, in particular if fully automatic verification is desired. Since the last iteration there is a new version of Frama-C which might have more options for interactive proofs than just resorting to the Coq interactive theorem prover. We feel that interactive theorem proving is a separate topic which would require using a different language than C for verification. This would reduce the applicability of the learning matter for our use-case.

Another challenge is finding a tool for abstract interpretation of C programs which works well with very simple domains. It would be possible to add such features to CBMC. There is an implementation of interval domains based on CBMC called intervalAI[7]. Unfortunately it is limited to the interval domain only, and also does not compile on current versions of gcc as is based on an older version of CBMC.

For a next iteration of this course this will likely change the sequence of topics and also shift the focus of the practical exercises from abstract interpretation to model-checking. Model-checking has the feature that a property that cannot be proven results in a counterexample which can be analyzed. This provided excellent direct feedback which allows for teaching about special and edge cases of fixed width vector based arithmetic of integers and IEEE 754 floating-point. Abstract interpretation will likely become a more theoretical topic, to be used on a whiteboard. It is a powerful technique in practice but without proper tool support it is difficult to teach in an applied setting.

References

1. Barrett, C., et al.: CVC4. In: Gopalakrishnan, G., Qadeer, S. (eds.) CAV 2011. LNCS, vol. 6806, pp. 171–177. Springer, Heidelberg (2011). https://doi.org/10.1007/978-3-642-22110-1_14
2. Baudin, P., Filliâtre, J.C., Marché, C., Monate, B., Moy, Y., Prevosto, V.: ACSL: ANSI C specification language. CEA-LIST, Saclay, France, Technical report v1 2 (2008)
3. Blanchard, A.: Introduction to C program proof with Frama-C and its WP plug-in. https://allan-blanchard.fr/frama-c-wp-tutorial.html
4. Bühler, D.: EVA, an evolved value analysis for Frama-C: structuring an abstract interpreter through value and state abstractions. Ph.D. thesis, Rennes 1 (2017)
5. Bühler, D., et al.: Eva-the evolved value analysis plug-in. https://frama-c.com/download/frama-c-eva-manual.pdf
6. Burghardt, J., Gerlach, J., Hartig, K., Pohl, H., Soto, J.: ACSL by example. DEVICE-SOFT project publication. Fraunhofer FIRST Institute (2010)
7. Clarke, E.M.: The birth of model checking. In: Grumberg, O., Veith, H. (eds.) 25 Years of Model Checking. LNCS, vol. 5000, pp. 1–26. Springer, Heidelberg (2008). https://doi.org/10.1007/978-3-540-69850-0_1
8. de Moura, L., Bjørner, N.: Z3: an efficient SMT solver. In: Ramakrishnan, C.R., Rehof, J. (eds.) TACAS 2008. LNCS, vol. 4963, pp. 337–340. Springer, Heidelberg (2008). https://doi.org/10.1007/978-3-540-78800-3_24
9. Filliâtre, J.C., Paskevich, A.: Why3 – where programs meet provers. In: Felleisen, M., Gardner, P. (eds.) Programming Languages and Systems, pp. 125–128. Springer, Heidelberg (2013)
10. Kirchner, F., Kosmatov, N., Prevosto, V., Signoles, J., Yakobowski, B.: Frama-C: a software analysis perspective. Formal Aspects Comput. 27(3), 573–609 (2015)
11. Kroening, D., Tautschnig, M.: CBMC – C bounded model checker. In: Ábrahám, E., Havelund, K. (eds.) TACAS 2014. LNCS, vol. 8413, pp. 389–391. Springer, Heidelberg (2014). https://doi.org/10.1007/978-3-642-54862-8_26

---

[7] https://github.com/sukrutrao/IntervalAI.

A Proposal for a Framework to Accompany Formal Methods Learning Tools

(Short Paper)

Norbert Hundeshagen[✉] and Martin Lange

Theoretical Computer Science/Formal Methods, University of Kassel,
Wilhelmshöher Allee 71, 34121 Kassel, Germany
{hundeshagen,martin.lange}@uni-kassel.de

Abstract. We propose a simple concept framework to accompany learning tools in theoretical computer science/formal methods in order to ease their integration into existing courses, balancing out technical issues against light-weight didactical questions.

Keywords: Theoretical computer science · Didactics of informatics · Software learning tools

1 Learning Tools in Theoretical Computer Science

Theoretical computer science (TCS) courses are usually quite formal and mathematical, and therefore often perceived to belong to the hardest subjects that students face. Failure and drop-out rates have always been high [12] with little indication that these rates would go down without active intervention [10]. Solutions to this problem range from the deployment of classic didactical methods (e.g. [10,15,16]), the use of adapted generalised tools, for instance using theorem provers for teaching logic [9], and the design, implementation and evaluation of specialised learning tools for particular subjects or tasks, e.g. [2,7,11].

This paper is concerned with development in the last of these categories, specifically seen from the perspective of theoretical computer scientists with rather little expertise in the field of didactics. We observe a growing interest and activity in the development of learning tools, as witnessed for instance by multiple venues like FMTea, ThEdu, FOMEO, and growing acceptance rates of papers on learning tools at non-educational conferences, c.f. [1,2,8,14]. Yet, in order to maximise learning effects, the development of such software tools needs to be embedded into a larger framework which is guided not only by technical but equally by didactical considerations, specialised to situations and issues commonly found in learning.

Modern learning tools generally make use of advanced technology to produce content and interactive user feedback, so that they can be – up to a certain degree – described under the term *intelligent tutoring systems* [6]. Intelligence is

© Springer Nature Switzerland AG 2021
J. F. Ferreira et al. (Eds.): FMTea 2021, LNCS 13122, pp. 35–42, 2021.
https://doi.org/10.1007/978-3-030-91550-6_3

often achieved using sophisticated formal methods (FM) like equivalence checks, SAT/SMT solving, theorem proving, model checking, semi-decision procedures, machine learning. These impact on didactical consideration which are rather specific to the area of TCS/FM. It is reasonable to assume that they have been implemented with a clear didactical purpose in mind, but this is often done by experts in TCS/FM, not necessarily in didactics, as the development of precise and efficient feedback technology for learning tools naturally requires expertise in the underlying technology.

We propose to accompany TCS/FM learning tools with a simple, yet clearly spelled-out didactical concept broken down to a collection of basic questions. It balances technical considerations with didactical aspects including addressed competence levels and feedback systems, thus encompassing the typical learning cycle. Such a framework would make the tools' developers provide a key element for the integration of such learning tools into an existing course and would help to bridge the gap between expertise in FM and comprehensive didactical theories. Moreover, it aids the comparability amongst learning tools, not necessarily to single out a best one but mainly to be able to quickly judge which tool is most suitable for which teaching situation. This would also enhance the effectiveness of the FM Courses Database[1] for instance by providing lecturers with information on typical didactic considerations, as prerequisites needed to deploy such tools, didactic limitations in using them, how learning outcomes are foreseen to be achieved, etc.

We give two examples of learning tool descriptions according to this framework. We conclude with remarks on continuing conceptual work accompanying the development of learning tools.

2   The Proposed Didactic Framework

Information that is rather obvious and normally given, like what subarea of TCS/FM the tool belongs to, what implementation technology has been used etc., should be complemented in such a framework with didactical considerations in order to facilitate a smooth integration into a course. For this to work, the concept description should particularly address the following issues.

What are the technical requirements for using the tool? This may restrict the answers to several other questions posed here, as it sets some key parameters for the use of the tool. Requirements may refer to the need to create user accounts, the availability of webserver, a database, the runtime environment of a programming language, other software, or the need to be run on a particular operating systems or within an educational platform etc.

What are the technical capabilities of the tool? The answer should be specific to the field of TCS as certain technical capabilities may show negative learning effects in particular fields. For instance, generating automated feedback on

---

[1] https://fme-teaching.github.io/#fm-courses.

an undecidable problem may impede students' understanding of the concept of decidability. For teachers it is vital to know the characteristics of the underlying technology (testing, bounded search, etc.) in order to be able to put the feedback system into context in students' eyes.

Furthermore, knowing the limits of the implemented technology to create learning effects also helps to judge their limits in creating such effects. For instance, when feedback is given in the form of counterexamples, it is useful to know the characteristics of methods used to produce them, i.e. taken from a fixed set, minimality, generated intelligently to address the student's individual learning problem, etc. Answers give key insight into didactical considerations such as: does the feedback system support the acquisition of certain competences, can feedback be misinterpreted, is it possible to mould an incorrect solution that passes all tests, etc.

The same considerations in didactical impacts also apply to tools that generate questions or exercises automatically. Technical capabilities that make content persistent or movable like the ability to save solution attempts to disc, to send or submit them, etc. can also influence didactical considerations likewise.

What does the course need to provide in order for the tool to be used? Specifically, what competences are the students assumed to already have acquired before they can start using the tool, for instance familiarity with a particular formal concept or the ability to carry out particular formal tasks. Larger tools which accompany an entire course rather than just target a specific competence may presuppose nothing in this respect.

Where in a course is the tool intended to be used? The standard learning model for courses in TCS still builds on the presentation of material in lectures, a short period of self-study to deepen and review understanding, plus set exercises/tutorial sessions. A clear recommendation on where to place the use of this tool in the line of activities making up a course is helpful, whether it is merely to be used in a lecture to visualise certain concepts, whether students are supposed to work on it autonomously perhaps after receiving a certain amount of tuition, or in a guided way during tutorials, whether the homework exercises can be run through this tool, etc. Multi-purpose in this sense is of course imaginable.

What competences at what level does the tool address? The answer has to be specific; it is not enough to know that tool X "helps students to learn logic." Instead the answer needs to name specific levels of competence, perhaps linking to standard syllabi in the respective area.

Even when there is no established formulation to name such levels yet it should be possible to sketch such levels. Often the analysis and categorisation of typical students' mistakes in exercises gives rise to a hierarchy of competence levels in a particular area, one that is quickly understood by fellow teachers. Listing standard exercises which students are supposed to be able to master after successfully learning with the underlying tool also provides valueable information here. Moreover, it can be useful to consider formulations of general competence

immediate feedback on impossible rule applications

here: attempt to apply the "implication-right-rule"

$$\frac{P(c){\to}P(f(c)) \;\Rightarrow\; \forall x \;.P(x){\to}P(f(f(x)))}{\forall x \;.P(x){\to}P(f(x)) \;\Rightarrow\; \forall x \;.P(x){\to}P(f(f(x)))}\;(\forall_\cup)$$

proof attempt: feedback system signals the provability of the sequent

$$\frac{\displaystyle \frac{\forall x\ \forall y\ .E(x,\,y){\to}x{=}f(y) \;\Rightarrow\; E(a,\,c){\wedge}E(b,\,c){\to}a{=}b}{\forall x\ \forall y\ .E(x,\,y){\to}x{=}f(y) \;\Rightarrow\; \forall z\ .E(a,\,z){\wedge}E(b,\,z){\to}a{=}b}\;(\forall_R)}{\displaystyle \frac{\forall x\ \forall y\ .E(x,\,y){\to}x{=}f(y) \;\Rightarrow\; \forall y\ \forall z\ .E(a,\,z){\wedge}E(y,\,z){\to}a{=}y}{\forall x\ \forall y\ .E(x,\,y){\to}x{=}f(y) \;\Rightarrow\; \forall x\ \forall y\ \forall z\ .E(x,\,z){\wedge}E(y,\,z){\to}x{=}y}\;(\forall_R)}\;(\forall_R)$$

proof attempt continued with unwise step: the provability status changed

$$\frac{\displaystyle \frac{\forall y\ .E(a,\,y){\to}a{=}f(y) \;\Rightarrow\; E(a,\,c){\wedge}E(b,\,c){\to}a{=}b}{\forall x\ \forall y\ .E(x,\,y){\to}x{=}f(y) \;\Rightarrow\; E(a,\,c){\wedge}E(b,\,c){\to}a{=}b}\;(\forall_L)}{\displaystyle \frac{\forall x\ \forall y\ .E(x,\,y){\to}x{=}f(y) \;\Rightarrow\; \forall z\ .E(a,\,z){\wedge}E(b,\,z){\to}a{=}b}{\displaystyle \frac{\forall x\ \forall y\ .E(x,\,y){\to}x{=}f(y) \;\Rightarrow\; \forall y\ \forall z\ .E(a,\,z){\wedge}E(y,\,z){\to}a{=}y}{\forall x\ \forall y\ .E(x,\,y){\to}x{=}f(y) \;\Rightarrow\; \forall x\ \forall y\ \forall z\ .E(x,\,z){\wedge}E(y,\,z){\to}x{=}y}\;(\forall_R)}\;(\forall_R)}\;(\forall_R)$$

Fig. 1. The feedback system in the Sequent Calculus Trainer.

hierarchies for entire study programmes, e.g. [3] as well as heavy-handed didactical considerations on how to build competence models for theoretical computer science [13].

What learning model does the tool follow? An answer would typically refer to standard behaviouristic, constructivistic, and cognitivistic learning models. Users of those tools, i.e. teachers integrating them into their courses, may not be familiar with such complete theories. It is therefore helpful to refer to specific methods advocated in such theories. Especially present feedback systems should be linked to detailed explanations on how learning is initiated, e.g. in an error-driven way, by immediate feedback on every user interaction, through simple questionnaires on the content, through feedback-loops, etc.

3   Two Exemplary Instances of the Proposed Framework

We exemplify the proposal of an accompanying framework using two specific tools: the Sequent Calculus Trainer [5] and DiMo [8]. The first one is – as the

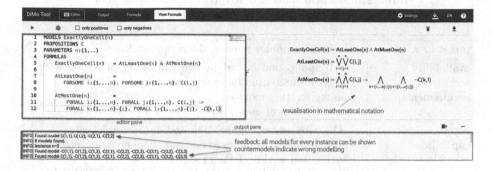

Fig. 2. Main view of the DiMo web frontend with highlighted feedback system.

Table 1. Didactical concept description for the Sequent Calculus Trainer.

The Sequent Calculus Trainer	
target area	– logic in computer science – formal proof systems (sequent calculus) – not useful for other proof systems like resolution etc.
availability, technical requirements	– free software (BSD-3 license) – written in Java, needs JRE 8.1 or higher – download: https://www.uni-kassel.de/eecs/fmv/ software/sequent-calculus-trainer
key technical capabilities	– presentation of formulas in mathematical notation – syntax highlighting – zoom-/scrollable proof display in tree shape – saving/loading/checking of proofs – validity checks for propositional and first-order logic (incomplete, uses SMT solver Z3) – limitations: feedback works best on shallow formulas and terms
provisions by the course	– introduction of the sequent calculus for first-order or propositional logic: proofs, rules, validity – not needed: soundness, completeness of the calculus
intended use	– (by teacher) presentation of proof construction, e.g. during lectures – (by student) (guided) solving of exercises on proving validity of formulas in sequent calculus during Tutorials and/or homework
learning model aspects	– immediate feedback for wrong use of concepts, see Fig. 1 – provision of hints to help students construct correct proofs – SMT solver checks possible next proof steps for feedback on successful use (traffic light system, see Fig. 1)
required competences	– familiarity with the syntax of first-order or prop. logic – helpful: ability to understand the meaning of formulas
addressed competences	– syntactically correct proving in a formal proof system – finding proofs for valid formulas – to a smaller degree: understanding the logical reason for invalidity/existence of counterexamples

Table 2. Didactical concept description for the DiMo tool.

DiMo	
target area	– logic in computer science, discrete modelling
availability, technical requirements	– web application hosted by the University of Kassel under https://dumbarton.fm.cs.uni-kassel.de – no registration required – no collection of user data
key technical capabilities	– IDE with syntax highlighting and auto-completion – presentation of formulas in mathematical notation, see Fig. 2 – saving and loading DiMo programs – satisfiability, validity, equivalence, model enumeration (uses SAT solver) – limitations: only integer parameter, computation timeout depending on formula size and/or instances of the parameters
provisions by the course	– syntax and semantics of propositional logic – definition of generalized logical operators $\bigwedge, \bigvee$ – short introduction of the core concepts of the DiMo-language (given in the manual)
intended use	– (by teacher) to exemplify semantical concepts and discrete modelling; for use as SAT solver interface – (by student) solving exercises on discrete modelling in tutorials and/or homeworks
learning model aspects	– immediate feedback on wrong syntax by syntax highlighting (see Fig. 2) and compiler messages – support of learning the connection between semantics and the modelled problem by output of formula models, see Fig. 2
required competences	– familiarity with the syntax of propositional logic
addressed competences	– understanding semantics of propositional logic – familiarity with satisfiability, validity, equivalence – using the above to solve "real-world"-problems – ability to form reduction to SAT

name suggests – a learning tool for formal proofs in First-Order Logic using Gentzen's sequent calculus which is, besides resolution and natural deduction, one of the proof calculi that is commonly taught in undergraduate courses on logic in computer science or presented in standard textbooks thereof, cf. [4].

The second one supports learning the competence of discrete modelling in propositional logic, which is essential in undergraduate studies in computer science.

The choice for these two tools to exemplify instances of the proposed framework is not made for any specific reason other than the fact that they are being developed by the Theoretical Computer Science/Formal Methods group at the University of Kassel. We would clearly be less competent to present the possibly unformulated didactical concepts underlining tools developed elsewhere.

We do not propose to use a specific format for such a framework at this point. Tool developers should of course have the freedom to choose how best to convey the information about the key didactical aspects of their tool. It is also not clear – certainly not at this point – whether there could be some format that is best suited for all purposes. Here we choose a semi-formal description in order to concisely present the key aspects in textual form.

Table 1 concisely lists information regarding the Sequent Calculus Trainer w.r.t. to the set of framework questions listed above. Similarly, this kind of information on the didactic use of DiMo is given in Table 2.

4 Conclusion

We propose to accompany the development of learning tools in formal methods by instances of a simple didactical framework that gives answers to some key questions about comparability, usability, usefulness and effectiveness of such tools to facilitate an easier integration into existing courses. The elements of this framework are given by key questions to be answered by the tools' developers.

We believe that the most effective learning tools will be created when knowledge and expertise from both formal methods and didactics is combined effectively in the design processes, and it seems like there is still potential to strengthen this kind of combination.

The framework proposed here should be simple enough so that theoretical computer scientists, in particular non-experts in didactics, are able to formulate basic properties pertaining to the didactic aspects surrounding their tools. In this way, didactics provides the means for transferability of formal methods learning tools between educational situations.

The idea that learning tools may be more transferable when shipped with spelled out didactical concepts is of course not restricted to the field of formal methods or theoretical computer science. Here we contained the field of interest in order to work on the basis of a clear and defined picture of recent developments and particular needs (like typical problems students face in this area) etc. This is not to say that other fields may not undergo similar considerations.

At last, the framework here is not considered to be completed. It should be discussed and engineered further under consideration of significant developments in the areas of didactics and formal methods. It also remains to be seen whether formalisation into a more stringent format would be beneficial.

References

1. Andersen, J.R., et al.: CAAL: concurrency workbench, Aalborg edition. In: Leucker, M., Rueda, C., Valencia, F.D. (eds.) ICTAC 2015. LNCS, vol. 9399, pp. 573–582. Springer, Cham (2015). https://doi.org/10.1007/978-3-319-25150-9_33

2. D'Antoni, L., Helfrich, M., Kretinsky, J., Ramneantu, E., Weininger, M.: Automata tutor v3. In: Lahiri, S.K., Wang, C. (eds.) CAV 2020. LNCS, vol. 12225, pp. 3–14. Springer, Cham (2020). https://doi.org/10.1007/978-3-030-53291-8_1
3. Desel, J., et al.: Empfehlungen für Bachelor- und Masterprogramme im Studienfach Informatik an Hochschulen. GI-Empfehlungen (2016)
4. Ebbinghaus, H.D., Flum, J., Thomas, W.: Mathematical Logic. Undergraduate Texts in Mathematics, 2nd edn. Springer, Berlin (1994). https://doi.org/10.1007/978-1-4757-2355-7
5. Ehle, A., Hundeshagen, N., Lange, M.: The sequent calculus trainer with automated reasoning - helping students to find proofs. In: Proceedings of 6th International Workshop on Theorem Proving Components for Educational Software, ThEdu 2017. EPTCS, vol. 267, pp. 19–37 (2017)
6. Freedman, R.: What is an intelligent tutoring system? Intelligence 11(3), 15–16 (2000)
7. Geck, G., Ljulin, A., Peter, S., Schmidt, J., Vehlken, F., Zeume, T.: Introduction to ILTIS: an interactive, web-based system for teaching logic. In: Proceedings of 23rd Annual ACM Conference on Innovation and Technology in Computer Science Education, ITiCSE 2018, pp. 141–146. ACM (2018)
8. Hundeshagen, N., Lange, M., Siebert, G.: DiMo – discrete modelling using propositional logic. In: Li, C.-M., Manyà, F. (eds.) SAT 2021. LNCS, vol. 12831, pp. 242–250. Springer, Cham (2021). https://doi.org/10.1007/978-3-030-80223-3_17
9. Knobelsdorf, M., Frede, C., Böhne, S., Kreitz, C.: Theorem provers as a learning tool in theory of computation. In: Proceedings of 2017 ACM Conference on International Computing Education Research, ICER 2017, pp. 83–92. ACM (2017)
10. Knobelsdorf, M., Kreitz, C., Böhne, S.: Teaching theoretical computer science using a cognitive apprenticeship approach. In: Proceedings of 45th ACM Technical Symposium on Computer Science Education, SIGCSE 2014. ACM (2014)
11. Rodger, S.H.: JFLAP: An Interactive Formal Languages and Automata Package. Jones and Bartlett (2006)
12. Ross, R., Grinder, M., Kim, S., Lutey, T.: Loving to learn theory: active learning modules for the theory of computing. SIGCSE Bull. 34(1), 371–375 (2002)
13. Schlüter, K., Brinda, T.: Characteristics and dimensions of a competence model of theoretical computer science in secondary education. In: Proceedings of 13th Annual SIGCSE Conference on Innovation and Technology in Computer Science Education, ITiCSE 2008, p. 367. ACM (2008)
14. Schwarzentruber, F.: Hintikka's world: agents with higher-order knowledge. In: Proceedings of 27th International Joint Conference on A.I., IJCAI 2018, pp. 5859–5861 (2018)
15. Sigman, S.: Engaging students in formal language theory and theory of computation. In: Proceedings of 38th SIGCSE Technical Symposium on Computer Science Education, SIGCSE 2007, pp. 450–453. ACM (2007)
16. Verma, R.M.: A visual and interactive automata theory course emphasizing breadth of automata. In: Proceedings of 10th Annual SIGCSE Conference on Innovation and Technology in Computer Science Education, ITiCSE 2005, pp. 325–329. ACM (2005)

Increasing Engagement with Interactive Visualization: Formal Methods as Serious Games

Eduard Kamburjan[1] and Lukas Grätz[2]

[1] University of Oslo, Oslo, Norway
eduard@ifi.uio.no
[2] Technische Universität Darmstadt, Darmstadt, Germany
lukas.graetz@tu-darmstadt.de

Abstract. We present a concept to increase the interactivity of formal methods courses. To do so, we discuss how formal methods can be seen as special serious games—a set of systems that is applied successfully in other educational contexts. To close the gap between the presented theory and its formalization or implementation, we take results from interactive visualization to develop a tool that empowers the students to deepen their knowledge about the presented theory in the same terms that are used in the lecture. The concept is not only based on experiences of the formal methods community, but also on studies and theories in the educational sciences. An implementation that is used in the exercise sessions of a course teaching proof calculi is available under https://kbar. app.

1 Introduction

Motivation. Teaching methods for formal methods have received renewed attention in recent years. Several workshops [6,9] and surveys [31,34] have identified numerous subject-specific challenges. One of the identified challenges is that formal methods tools are not designed to be used for teaching: they have a steep learning curve and give little feedback [34]. As such, they are at most useful to teach how to *solve* a problem using the formal method but give little support to teach their internal *concepts*. Two further challenges worsen the situation: For one, formal methods are dry and many students feel intimidated by mathematical expressions to the point of mathematical anxiety [28]. For another, the presentation of formal methods in lectures, their implementation and their presentation in textbooks is highly non-uniform.

It is not possible to use the tool to manipulate and explore the presented material using the concepts *as presented in the lectures*. This results in low student engagement in formal method courses, as the gap between tool and concepts discourages the student.

Overview. In this work, we use the structural similarities between formal methods and serious games to develop a concept to increase student engagement

© Springer Nature Switzerland AG 2021
J. F. Ferreira et al. (Eds.): FMTea 2021, LNCS 13122, pp. 43–59, 2021.
https://doi.org/10.1007/978-3-030-91550-6_4

in formal methods courses. Serious games have been shown to have a positive effect on student engagement [32] and our concept aims to carry over this effect to formal methods. To see a formal method as a serious game, one needs a visualization with the following characteristics: (i) It is interactive. (ii) It visualizes the formal methods consistently with the concepts and notations in the theory part of the course.

The added value of interactive visualization is that it communicates the mental model of the expert more faithfully [18]. In contrast, formal method tools that focus on applications, need a different mental model, and thus, fail to increase student engagement. The interactivity here is crucial: non-interactive visualizations have only a small effect on student performance [19]. Note that in this concept the formal method itself is the serious game; it is *neither* an application of a formal method to games *nor* a gamification of existing tools.

This work discusses a teaching concept on how the view on formal methods as serious games can be used to develop and apply an interactive visualization tool in a formal method course. In our concept, the interactive visualization tool is used (a) as a self-study help for students and (b) in the exercise session of a course. In early learning stages, a student internalizes new concepts. Here, using the tool as a self-study can help students to explore the concepts on their own terms and on their own speed. As the tool catches mistakes and provides feedback why certain operations are not applicable, students do not reinforce misconceptions by reapplying mistakes. In contrast to pen-and-paper exercises, where they would have to wait several days for the teacher's feedback.

Proof Calculi. We present our approach to develop a tool for teaching proof calculi, e.g., different tableaux variants, based on the authors' experience with the exercises of a course on automatic theorem proving. Consistently with the identified challenges, we observe that it is challenging for students to adopt the taught calculi to prove something on new examples. We conjecture that one of the reasons, for example in tableaux methods, is that the lecture presents the proof calculus visualized as a tree, creating a cognitive gap to the formalization (where the tree is implicit behind mathematical notation) and the implementations.

We present the KalkulierbaR tool, which allows the student to build and change proofs in several variants of tableaux, resolution, DPLL and sequent calculi. The tool is a serious game, similar to puzzle games, where the proof rules are possible steps in the game. The goal is to close the proof.

Contributions. The main contributions of this work are (a) a concept to view formal methods as serious games to teach formal method theory, and (b) the KalkulierbaR tool, designed to be part of this concept to teach several proof calculi. The tool is designed by the authors, the implementation of KalkulierbaR was done together with two groups of students, which were implementing user stories given by the authors as part of a mandatory lab. The format is explained in detail in Sec. 4. It includes several variants of tableau, resolution, DPLL and the sequent calculus. A live instance runs under https://kbar.app and the source is available under https://github.com/kalkulierbar/kalkulierbar.

Structure. In Sect. 2 we present background on serious games and interactive visualization in education, as well as challenges specific to formal methods and related work. Sect. 3 applies these ideas for formal methods to present KalkulierbaR. Sect. 4 discusses our experiences and Sect. 5 concludes.

2   Background and Related Work

We discuss the background for serious games, interactive visualizations and student engagement from a formal method perspective.

We follow Roggenbach et al. [29] and understand a "formal method" as (a) a set of syntax with some (b) semantics and (c) a set of rules operating on the syntax. This includes logics with proof calculi. Under a logic we understand an abstract logic [14], i.e., a triple of (a) a set of sentences as syntax, (b) a set of models and a satisfiability relation with certain properties as semantics, and (c) a proof calculus that consists of operations on the syntax as rules.

Interactive Visualizations. Algorithm and program visualization has a long history in computer science education and is documented dating back to the beginning of the 1980s[1]. It seems intuitive that visualizations help to engage the student and increase the accessibility of taught material, and several theories on why this is indeed the case have been put forward. For example, the epistemic fidelity theory of Hundhausen argues that algorithm visualizations *"provide a faithful account (i.e., one with high epistemic fidelity) of an algorithm's execution in terms of an algorithm expert's mental model"* [18].

However, a meta-study of Hundhausen et al. [19] recognizes that providing algorithm visualization tools alone does *not* increase student performance:

> *"With few exceptions, we found that studies in which students merely viewed visualizations did not demonstrate significant learning advantages over students who used conventional learning materials"*.

Instead, increased student performance can be attributed to *interactive* elements of the visualization tool, such as constructing own input sets, programming tasks, answering questions and building own visualizations [24]. Hundhausen et al. also find that tools designed for active learning and based on *cognitive constructivism* have more significant results in increasing student performance. Naps et al. argue that *"visualization technology, no matter how well it is designed, is of little educational value unless it engages learners in an active learning activity* [24]". They present an engagement taxonomy of possible involvement: (1) no viewing, (2) viewing, (3) responding, (4) changing, (5) constructing and (6) presenting. Subsequently, this taxonomy was extended by Myller et al. [23] to include more fine-grained levels.

We observe that formal methods are easy to adapt to interactive visualization for multiple forms of involvement in the engagement taxonomy (2), (3) and (4).

---

[1] The most prominent artifact from these early approaches is the short film Sorting Out Sorting [4]. For a historic overview, we refer to Baecker [5].

There is a major difference between the algorithms taught in typical algorithm and data structure courses, which form the basis for the above studies, and formal methods: Many formal methods are non-deterministic (without a strategy for rule selection) and as such enable even more interaction than algorithm visualization. However, we stress that the reason for increased student performance is that the visualization is used to provide and manipulate a certain mental model that is taught in the lecture. As such, application-oriented tools are not suitable for this task—there is a gap between how a method is explained in its pure form and how it is implemented. Even systems with a focus on user interactions, such as the KeY tool [1], work on an extension of the pure method and for novices, it is not easy to distinguish the parts which stem from the pure method and which stem from application needs.

Learning with Serious Games. A similar approach to interaction in learning are (educational) serious games. There is no agreement on the exact definition of a serious game[2], and for the rest of this paper, we commit to using the one put forward by Wouters et al. [32]: interactive systems based on a set of agreed rules and constraints, directed toward a goal, which provide feedback to the player to enable monitoring the progress towards the goal.

Serious games are *not* necessarily about gamification: Serious games are game-like programs for education, while gamification is about game-like mechanics in non-game contexts [21], e.g., awards for participation in online discussions. Nonetheless, serious games must be embedded in context: a meta-study of Wouters et al. [32] concludes that while serious games can lead to better learning and retention, the students do *not* feel more motivated by them.

We discuss the exact connection with serious games in Sect. 3.1. For now, it suffices to remark that we can interpret the set of syntactic rules (of the formal method) as the goal-directed rules of a serious game. The goal is, for example, to close a proof. For a formal method to become a serious game, we have to add visualization, interaction, and feedback.

Related Work in Formal Methods. Cerone et al. [7] discuss specific challenges for teaching formal methods based on a recent workshop. While most of their discussion focuses on the role of formal methods in a curriculum for software engineers, some of the authors also mention the role of games in their teaching. However, they use games as case studies to introduce formal methods, e.g., to have an intuitive set of rules that needs to be modeled *using* the formal method. Tools developed on this idea, like the `FormalZ` tool of Prasetya et al. [27], are successfully applying gamification to the course, but do not see the formal methods as the serious game itself.

Another raised point is that the tools are not suitable due to confusing error messages and interface. This coincides with the epistemic fidelity theory: these tools do not provide a faithful account of the mental representation of the *teacher*

---

[2] Defining a game is notoriously challenging. For a recent survey from the view of electronic games, we refer to Arjoranta [2].

Table 1. Selected tools for proof calculi and criteria for serious games.

Tool	Rules	Interactive	Visual	Feedback
Sequent calculus trainer [10, 11]	Sequent calculus	✓	✓	✓
Panda [15]	Natural deduction	✓	✓	?
Easyprove [22]	Pen-and-paper math	✓	✗	✓
CalcCheck [20]	Pen-and-paper math	Partial	✗	✓
WinKE [8, 12]	Tableaux calculus	✓	✓	?
KalkulierbaR	Multiple calculi	✓	✓	✓

for theoretical concepts, as they are based on the mental representation used for their *application by experts*. Farell and Wu discuss in a recent experience report [13] also the problems of relating tools and theory due to the disconnect of concepts as taught in the course and concepts as used in the tool. We see their experiences as representative for several studies reported in the aforementioned white paper of Cerone et al. [7] as well as other surveys [31,34], which also identify a lack of tool support and visualization as challenges for formal methods.

There are numerous visual interfaces for single proof calculi (e.g., [8,10,15, 22]). Additionally, we present KalkulierbaR in Sect. 3, a tool that covers multiple calculi and is explicitly designed following the didactical theories behind interactive visualization and serious games. Not all of these tools support all features of a serious game, see Table 1. For example, CalcCheck and Easyprove were intentionally designed for text-based math proofs and not for visual proofs. The Sequent Calculus Trainer is an interactive visualization of the sequent calculus, following a didactical motivation [11]. Since it also provides interactive feedback, it fits our perception of a serious game—although not explicitly designed as a such.

As we will discuss in the next section, interactive provers can be seen as serious games, but not necessarily as *interactive* visualizations. This is, for example, the case for Isabelle and Coq, which have a *textual* interface. Textual interfaces are suitable for different teaching approaches. For example, the CalcCheck tool [20] aims to give an interface that is as near as possible to the notation (and language) in the used textbook. Similar ideas are used by Pierce et al. [26] in a series of books that are executable Coq scripts. The exercises in these books can be seen as serious games (without interactive visualization), but this connection is not made explicit.

Note that we focus here on the aspect of teaching concepts – powerful tools are still necessary in teaching if formal methods are taught in an applied program. Ölveczky [25] takes a different approach to this and argues that Maude is a tool that can bridge the gap between theory and application in one tool, as its formalism, a rewriting logic, is similar to functional programming. As such, Maude relies on prior knowledge of the student with another similar formalism, which is not possible for no-programming based formal methods. It is an example of a formal method where the gap between concepts and tool is small, in this work we focus on teaching methods for formal methods with a bigger gap.

3   Proof Calculi as Serious Games

Based on the principles discussed in the previous section, we suggest teaching methods to assist lectures, exercises and self-study with interactive visualization. We present KalkulierbaR, an implementation of this concept for proof calculi.

3.1   Formal Methods Are Serious Games

Our key observation is that formal methods are serious games, if an interactive visualization is provided. We remind that a serious game is an interactive system with a set of rules and constraints directed towards a certain goal, that provides feedback to monitor the progress towards the goal.

For formal methods, there is a set of agreed rules and constraints (operating on syntax) and a clear goal (closing the proof or reaching/avoiding some state). For example, for proof calculi, the goal of the game is to close the proof using a fixed set of proof rules. For, say, model checking of liveness properties in automata, the goal is to find a path to some location. Compared to other games, they are near to puzzle or tile-matching games.[3] Most notably, formal methods are single-player games and do not have an opponent. While they have a clear winning condition (reaching the goal) they do not necessarily have a clear losing condition. Such a condition is not necessary for a system to be a game.

As we see, formal methods merely lack interaction and feedback. This must be provided by an interactive visualization of the formal method. This means that even without gamification efforts, formal methods with interactive visualizations are serious games. As we have seen in the previous section, serious games and interactive visualization do not automatically increase student engagement or retention. Instead, they must be integrated into the course in a way that they are similar to the mental representation of taught concepts. We give a concept to do so in the rest of this section.

3.2   Teaching Methods

Our concept is to use an interactive visualization tool in a formal method course as a consistent help for the student. At the core, the tool allows the student to work with a formal method as it is taught in the lecture and textbooks. In particular, we aim to use the same syntax and visualizations for rules as already given in the lecture. Before we introduce our implementation, we describe where and how the tool is used in the course, with proof calculi as a guiding example.

We assume a tool that (1) visualizes the current state of a formal method (e.g., a proof tree or program configuration) (2) permits selection of parts of the state (e.g., a proof node), (3) displays a list of possible rules to apply and (4) provides detailed feedback if a rule is selected that cannot be applied.

---

[3] For instance, Sudoku, Tetris or Candy Crush. The similarity to such puzzles goes beyond the definition: Candy Crush and other three-matching games have been shown to be NP-complete despite their simple rule sets [16].

While optional, we also assume that the tool (5) permits reverting steps for backtracking.

Lectures. The tool can be used to increase the interactivity of a lecture by executing the taught formal method in the plenum. For example, a teacher performs a proof but asks the students how to proceed in each step. A student can answer with a rule and a proof node. In an online live session, the answers can be typed in a chat. As a variant, for lectures with 20 students and more, anonymous polling software could be used to find the next step by majority vote.

Advantages of using an interactive proof tool in the lecture:

Flexibility. Contrary to slides, a teacher can react to unforeseen answers of the students and explore alternative strategies suggested by the students. Contrary to (digital) blackboards, we do not need to reserve space beforehand: Scaling and arranging a proof (tree) is done automatically by the software. Similarly, it is less time-consuming to perform proof steps, further facilitated by reverting or undoing steps. Student answers also provide valuable feedback for the teacher to assess the learning progress.

Student Engagement. The tool fosters category 3 in the engagement taxonomy (responding), as it can be used to ask the students questions (e.g., "what will be the result of this step"). Tool support also helps using category 4 (changing), where the students can influence the next steps of the formal method.[4]

Low Threshold for Participation. The method using an interactive proof tool lowers the participation threshold in multiple ways: Since students also have access to the tool, they can protect themselves before answering by performing the steps beforehand. Furthermore, students (otherwise anxious to participate) are activated, since they only need to follow the rules of a calculus. The answers are often a few letters only, which could be typed (depending on the lecture format) in an anonymous chat.

Reproducibility. Students can reproduce proofs using the same tool.

Self-study. After an aspect of a formal method is introduced in the lecture, students not only have the possibility to reproduce the examples, but also to explore different strategies and examples on the students' own time and terms. In our proof calculus setting, a student selects a node in the proof and an appropriate rule. The proof is drawn by the software and feedback is given, whenever the student tries to apply a rule in a wrong way. To achieve the final goal, a specific order of rule applications may be necessary: By using *backtracking* in the form of undoing proof steps, the student may fix this order.

In particular, the student can (a) stay within the conceptualization introduced in the lecture, (b) get *step-wise feedback* on erroneous input (e.g., trying to apply a rule that does not match the situation) in terms of the very same conceptualization and automatically. It is neither necessary to learn new syntax

---

[4] In terms of AV, changing is mostly used to provide new inputs to the algorithms. Formal methods have more flexibility in possible input.

or different visualizations as used by tools aimed for applications, nor necessary to connect error messages from application tools to the basic concepts of the formal method. Ehle et al. [10] have observed that syntactically wrong rule instantiations are a major source of errors.

Exercises. Exercises come with a variety of different tasks and are an essential part in learning formal methods. In exercise/lab sessions, the student is given multiple tasks to apply the content of the lecture. In the first parts of the course, this is done on a conceptual level. Some tasks can be solved interactively in a calculus as described above. For example, when we give students a formula that needs to be proven using a calculus. The advantages of interactive visualization in this context are the same as in the lecture setting described above: Both pen-and-paper solutions and interactive visualization solve the same problem, but by using the software, the students benefit from step-wise and early feedback.

Interactive visualization is mainly suited for exercises on reasoning *within* a formal method (formal proofs in a calculus, algorithmic proof procedures, etc.). Exercises on meta-level properties *of* a formal method do not benefit directly (like soundness of a calculus).

Labs. There is a difference between using and extending a tool. Students benefit from both. On one hand, *using* allows one to perform steps in the formal method with step-wise feedback, as described above. On the other hand, *extending* engages students even further: Category 5 of the engagement taxonomy is to construct new visualizations. We suggest the following labs:

Strategies. We assume a formal method to be non-deterministic. However, it likely has a reasonable strategy for rule selection. Such strategies can be implemented in the interactive visualization tool. Additionally, it may be used to visualize auxiliary structures used for the strategies (such as "set of support" in resolution).

Extensions. An interactive visualization tool could be a basis for the implementation of additional rules or alternative rule sets. Extensions include the calculus itself, the syntax, or the graphical view.

Exams. Interactive visualization tools have limited applications in exam situations. Students are generally expected to perform pen-and-paper proofs by themselves and need to demonstrate that they understood the system *without* a program to guide them. However, students certainly benefit from the interactive visualization tool when preparing for the exam by self-study, as described above.

There are exam situations when pen-and-paper are difficult to organize, e.g., in (open-book) online written exams or online oral exams. In these situations, students may use the interactive proof tool. It should be carefully monitored that (1) such exams are still being fair to all students and (2) students' performance is independent and not bound to a particular proof assistance tool. Furthermore, we should respect students' privacy and the in-homogeneity of environment and infrastructure on the students side.

Table 2. Calculi implemented in `KalkulierbaR`, listed with supported variants and logics. PL denotes propositional logic, FO denotes first-order logic.

Calculus	Variants	Logics	Model generation
(Clausal) tableaux	Regular, (strongly) connected	PL, FO	×
Non-clausal tableaux		FO, Modal	×
Resolution	Hyper resolution	PL, FO	×
Sequent calculus		PL, FO	×
DPLL		PL	✓

3.3 KalkulierbaR

`KalkulierbaR` is an interactive formal proof tool designed to support teaching proof calculi following the concept described above. We refrain to introduce the calculi in detail and instead illustrate the usage of `KalkulierbaR` by example. Table 2 shows the implemented calculi and their variants. `KalkulierbaR` is implemented as a web application, consisting of a frontend (written in JavaScript) for the interface and a backend (written in Kotlin) for the state.

`KalkulierbaR` uses a responsive design and can be used with a touchscreen (on a smartphone) as well as with a pointer device (desktop or laptop). Thus, we rely only on standard web technology and no installation is necessary, reducing the threshold for the student to use the tool. The backend can be installed locally and includes a build system that downloads all dependencies.

Overview. Initially, the students select a calculus with a suitable logic and set the parameters, for example, the weakly connected tableaux variant. Additionally, there are some additional settings, such as backtracking: the ability to undo steps. Then a formula is entered (a clause-set or a sequent, depending on the chosen logic and calculus). At this point, one can set optional parameters. Some parameters control details of the calculus, like restricting tableaux to weakly connected tableaux, while others influence the overall workflow, such as backtracking. In the serious game view, this corresponds to adjusting the rule set.

The input is then parsed and sent to the backend server, where logic and calculi are implemented. If parsing fails, error messages are provided. Before the proof is started, the backend server might perform some normalization steps.

Once the calculus is selected and a formula is entered, one can start playing by applying rules of the respective calculus. When a proof is shown for the first time, a tutorial appears. The tutorial can be reopened using the `help` button on the screen. Usually, deduction steps in the calculi consist of selecting one or two formulas and a rule to apply. The rules are either in the lower right or on the left. Sometimes, a rule might request additional parameters in a pop-up window.

Once a proof is finished we can use the `check` button to verify our result. Properties of the proof are displayed in the message box.

Example. We give an example to demonstrate `KalkulierbaR`. Our example is in spirit of Smullyan [30], Chapter XIV, and gives students a more motivating exercise then an abstract formula. It also demonstrates the full range of the exercise, from situation, over logical modeling to the use of calculi for solving.

"Once, a parent went grocery shopping with their child. Most notably, they bought yogurt and a chocolate egg. The parent says: 'My dearest child, you have done so well today. I would like to reward you, only you have to say the truth in whatever statement. If your statement is true, then I will reward you with the yogurt, but if the statement is false, I will not give it to you.' Now it so happened that the child wanted to have the egg and not the yogurt! The child is clever and makes the following statement: 'You will give me neither the egg nor the yogurt'".

This statement forces the parent to give the egg to his child, as only this outcome makes the demanded statement by the child false without breaking the promise of the parent. We could check that this produces the desired outcome by examining all possible combinations manually. But by applying formal methods, we can do this more systematically.

First, we need to formalize the statement in logic. Our formalization is given in Table 3 and uses two propositional variables e and y.

Table 3. Formalization of Smullyan's puzzle.

Proposition	Meaning
y	"The parent gives the yogurt to the child"
e	"The parent gives the chocolate egg to the child"
$\neg(y \vee e)$	Statement by the child: "You will not give me neither the yogurt nor the egg"
$\neg(y \vee e) \rightarrow y$	First proposition by the parent: "If the statement is true, then I will give you the yogurt"
$\neg\neg(y \vee e) \rightarrow \neg y$	Second proposition by the parent: "If the statement is false, then I will not give it to you"

There are two parts in the verification of the puzzle's solutions. We show that the desired conclusion follows, assuming the parent keeps to true to their word. Furthermore, we have to show consistency of the premises, i.e., we check that it is possible to actually keep the parents' word. We solve the first part using a proof calculus and the second part by checking a satisfiable model.

We show that the parent gives the yogurt to the child, whenever both propositions by the parent are true. To do so, we use the sequent calculus, where premises are on the left and the conclusion on the right of the turnstile $\vdash$:

$$\neg(y \vee e) \rightarrow y, \quad \neg\neg(y \vee e) \rightarrow \neg y \vdash e \tag{1}$$

We may enter the sequent in the respective ASCII-notation, as shown on the left in Fig. 1. Once we start the proof, the input sequent is displayed in the usual notation at the bottom. Now we can try applying rules like "impLeft" on $\neg\neg(y \vee e) \rightarrow \neg y$. The proof tree after this step is shown at the top of Fig. 2. Whenever the student tries to apply a rule on a formula where it is not possible, the exact cause is displayed. There are multiple ways to finish the sequent proof, one of them is shown at the bottom of Fig. 2.

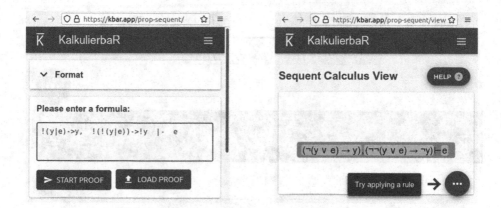

Fig. 1. Starting a sequent proof with KalkulierbaR.

Model Generation. For DPLL, KalkulierbaR supports checking whether a model satisfies the original formula. For the above example, a resulting model is given by $\neg y$ and e, i.e., the parent gives the egg but not the yogurt. Due to the course contents (and not due to theoretical reasons), model generations had been restricted to DPLL.

Variants and Layout. As discussed, KalkulierbaR allows the user to select a variant of the used calculus. For example, the user may use *regular* tableaux, where no literal may occur twice on a branch. If such a restriction is violated, the user is informed which of the variants is not followed correctly and where. Once the proof is closed, the user is also informed which variants could have been activated. For example, the left side of Fig. 3 shows a closed proof with the corresponding message. Beyond feedback on erroneous rule applications, KalkulierbaR also has a button that explains all currently available rules and shows animations to illustrate the rule with an example.

Tableaux is the main focus in the course and multiple more advanced calculi are implemented to show variants of tableaux beyond the clausal-based system for standard first-order logic and propositional logic: KalkulierbaR provides classical non-clausal tableaux and signed modal tableaux [33]. Just to give an idea of other calculi: On the right of Fig. 3 is a small modal logic proof for the basic modal axiom K.

The layout of DPLL, sequent calculus and tableaux is fixed. For resolution, as no specific layout is used in the lecture, we let the user switch between two

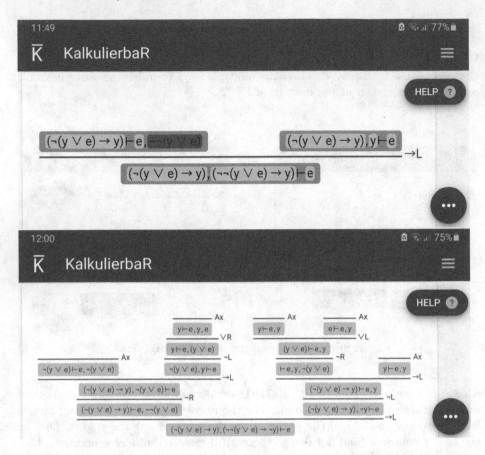

Fig. 2. A sequent proof with `KalkulierbaR` on a mobile phone display.

possible layouts. One where the clauses are arranged in a circle and one where the clauses are arranged in a grid. In the circle, whenever a clause is selected, the clauses are rearranged such that possible resolution partners are near the selected clause.

Backend. The backend permits to hide calculi and variants from the user through an admin interface. This is used to synchronize the course and the calculi offered by the webtool. The backend also automatically translates user input into conjunctive normal form, if the user selects a clausal calculus. Alternatively, the user may enter a set of clauses directly.

Additionally, high-score tables can be activated for certain calculi to compare properties with proofs from other users. After the proof is checked, the high score table appears and the user can enter a name to save store the result. We stress that this is *not* part of the serious game concept. It is a *gamification* approach that is orthogonal to the formal-method-as-serious-game view.

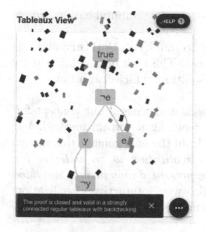
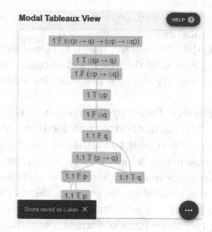

Fig. 3. A clausal tableau proof and a modal tableau proof.

4 Discussion

Application. The original development of KalkulierbaR started in winter 2019, based on experiences with the course on "Automated Theorem Proving" at TU Darmstadt. When the course was repeated in summer 2020, we had the opportunity to use KalkulierbaR in a course with 15 regularly attending students in the lecture.

As the COVID-19 pandemic forced us to change to a virtual setting, our tool was used differently as originally planned. Similar to other courses we observed that the short-term, unplanned switch to a live streaming format of courses negatively affected student interaction and engagement [17], making it difficult to compare its effects to previous iterations. There was no evaluation on the participants of the lecture in connection with the use of KalkulierbaR. Such an evaluation could give detailed feedback on how the usage of KalkulierbaR affected the learning process. We plan to make evaluations when the lecture is held again.

Exercises. KalkulierbaR was mainly used for exercises, where the exercise sheets could be solved using KalkulierbaR instead of pen-and-paper as in the previous years. The solution discussion took place in dedicated sessions. Following the structure of previous iterations of this course, the students were not required to submit their solutions. 4 out of 7 exercises were suited KalkulierbaR and the solutions were partially presented using KalkulierbaR.

Lecture. The course had 7 lectures on the concepts of tableaux, resolution and DPLL. For the respective calculi, KalkulierbaR was used as described in Sect. 3.2. As mentioned, the course suffered from the widely observed negative effect of the pandemic negatively on student engagement. Nevertheless, students actively participated using the text message function, both when

asked and on self-initiative by asking questions for comprehension, when
KalkulierbaR was used.

Labs. We offered one additional exercise sheet to modify KalkulierbaR, thus
realizing the *construct* category of engagement. The task of this lab was to
modify the rule used for first-order hyper resolution, which had a bug.

Further Forms of Involvement. The previous section discussed our concept for
a formal method course focused on an expert teaching a certain method. We
could extend it to further forms of involvement from the engagement taxonomy:
presenting, which is defined as *"presenting a visualization to an audience for
feedback and discussion. The visualizations to be presented may or may not have
been created by the learners themselves* [24]*"*. This way, interactive visualization
can also be used in a seminar setting where students read up, implement and
present variants of formal methods.

KalkulierbaR itself was implemented in two mandatory student lab (*"Bachelor Lab"*) which simulates an industrial development environment using agile
practices. The project lead, in our case the authors, has an already designed
application concept, that needs to be implemented and give the students user
studies in regular meetings, which they implement and get approved by the
project lead. We used KalkulierbaR in two consecutive such labs, where the
code from the first lab was extended in the second one. Most of the students had
not taken the ATP course before. These labs were educational tasks in themselves, realizing the "constructing" aspect in the engagement taxonomy. The
design of KalkulierbaR was not part of the students' work.

KalkulierbaR has a modular structure that separates the logical operations
in the backend from their visual representation in the frontend. As such, a possible lab would be to implement the proof strategies to make the system automatic,
such that the visualization can be used to examine the intermediate state without
modification. Modifying the visualization allows one also to explore the internal
state of the strategy. For example, implementing resolution with set-of-support
(SOS), requires adding the SOS clauses to the visual interface.

Furthermore, KalkulierbaR implements several calculi and can be extended
to support more. It permits a contrasting approach to teach proof multiple calculi
in a uniform interface. Single-calculus tools [8,10,15,22] cannot be used so.

On Generalization. The presented connection of formal methods with interactive
visualizations as serious games is independent of the chosen implementation for
proof calculi and we conjecture that the concept in Sect. 3.2 can be used for
any formal method. While there is an overhead to develop a tool specifically for
teaching one formal method, we deem it acceptable for the following reasons: As
the aim is to provide a serious game form of the method *as taught*, such tools
are more reusable by other lecturers than tools used for the application. It is
not necessary to have a teaching companion for an advanced tool and keep the
teaching tool up-to-date with the application-oriented tool.

That the teaching tool requires less maintenance is an important practical
point: Lack of maintenance is of the reasons for the growing disconnect between

KeY and its teaching companion[5], KeY-Hoare, which is based on KeY 1.6, while the current release version of KeY is 2.8 and includes usability improvements that cannot be used in KeY-Hoare. The development of the tool can be integrated with student projects to further increase engagement in an applied manner.

The serious games teaching companion described in this work can be complemented by application-oriented tools if the course structure includes such tools. In our course, we had some case studies using theorem provers as application-oriented tools. There is no redundancy—the teaching companion may be used to introduce the concepts and describe them in their pure form, while application-oriented tools can focus in later parts of the course on bigger case studies.

5 Conclusion

This work establishes a firm connection between formal methods on one side and interactive visualization and serious games on the other side: The formal method itself is a serious game, where the rules of the formal method are the rules of the game. To control the rules and get suitable feedback, the user needs an interactive visualization tool fitted for the formal method.

Using this connection, and further theory from educational sciences, such as the engagement taxonomy, we present a teaching concept tailored to the challenges of formal methods, in particular, the notoriously novice-unfriendly tools. The main goal is to increase student engagement in the theoretical parts of a course, by providing a specific teaching tool that helps to learn the concepts before applying them in the application-oriented tool.

Future Work. We plan to use KalkulierbaR in a more mainstream course to be able to perform a quantitative study on its effects on student engagement. To our best knowledge, there are no recent studies on interactive visualization after the advent of smartphones and their mass use by students. It is worth investigating whether this has an effect on how students react to interactive visualization.

We plan to integrate additional output formats, such as LATEX, and integration into a wider tool for online teaching that builds on KalkulierbaR for grading exercises and interactions with students through quizzes and chats, two tools that were also shown to increase engagement in an online setting. Finally, we consider using an additional module to input proof rules, e.g., MUltlog [3], to use KalkulierbaR in a setting with a more volatile treatment of proof calculi.

Acknowledgments. The authors thank Daniel Drodt, Julius Henk, Mirko Hirsch, Tim Kilb, and Nils Rollshausen, as well as Nils Elze, Lars Hoffmann, Enrico Martin, Henrik Metternich, and Ashim Siwakoti, who helped to implement KalkulierbaR during two student labs. The lecture part of the described course was hold by Reiner Hähnle and Richard Bubel. This work was partially supported by the Research Council of Norway via *SIRIUS* (237898) and *PeTWIN* (294600).

---

[5] KeY-Java is used to teach Java verification directly, not the general setting.

References

1. Ahrendt, W., Beckert, B., Bubel, R., Hähnle, R., Schmitt, P.H., Ulbrich, M. (eds.): Deductive Software Verification - The KeY Book - From Theory to Practice. LNCS, vol. 10001. Springer, Heidelberg (2016). https://doi.org/10.1007/978-3-319-49812-6
2. Arjoranta, J.: How to define games and why we need to. Comput. Games J. **8**(3), 109–120 (2019). https://doi.org/10.1007/s40869-019-00080-6
3. Baaz, M., Fermüller, C.G., Salzer, G., Zach, R.: MUltlog 1.0: towards an expert system for many-valued logics. In: McRobbie, M.A., Slaney, J.K. (eds.) CADE 1996. LNCS, vol. 1104, pp. 226–230. Springer, Heidelberg (1996). https://doi.org/10.1007/3-540-61511-3_84
4. Baecker, R. Sorting out sorting. Educational film (1980)
5. Baecker, R.: Sorting out sorting: a case study of software visualization for teaching computer science. In: Software Visualization: Programming as a Multimedia Experience, pp. 369–382 (1998)
6. Cerone, A., Roggenbach, M. (eds.) Formal Methods - Fun for Everybody, Revised Selected Papers. CCIS, vol. 1301. Springer, Heidelberg (2019). https://doi.org/10.1007/978-3-030-71374-4
7. Cerone, A., et al.: Rooting formal methods within higher education curricula for computer science and software engineering - a white paper. White paper (2020). https://arxiv.org/abs/2010.05708
8. D'Agostino, M., Mondadori, M., Endriss, U., Gabbay, D., Pitt, J.: WinKE: a pedagogic tool for teaching logic and reasoning. In: Goettl, B.P., Halff, H.M., Redfield, C.L., Shute, V.J. (eds.) ITS 1998. LNCS, vol. 1452, p. 605. Springer, Heidelberg (1998). https://doi.org/10.1007/3-540-68716-5_69
9. Dongol, B., Petre, L., Smith, G. (eds.): Formal Methods Teaching. LNCS, vol. 11758. Springer, Heidelberg (2019). https://doi.org/10.1007/978-3-030-32441-4
10. Ehle, A., Hundeshagen, N., Lange, M.: The sequent calculus trainer - helping students to correctly construct proofs. In: 4th International Conference on Tools for Teaching Logic TTL. abs/1507.03666 (2015)
11. Ehle, A., Hundeshagen, N., Lange, M.: The sequent calculus trainer with automated reasoning - helping students to find proofs. In: Quaresma, P., Neuper, W. (eds.) 6th International Workshop on Theorem Proving Components for Educational Software. EPTCS, vol. 267pp. 19–37 (2017). https://doi.org/10.4204/EPTCS.267.2
12. Endriss, U.: An interactive theorem proving assistant. In: Murray, N.V. (ed.) TABLEAUX 1999. LNCS (LNAI), vol. 1617, pp. 308–313. Springer, Heidelberg (1999). https://doi.org/10.1007/3-540-48754-9_26
13. Farrell, M., Wu, H.: When the student becomes the teacher. In: Cerone, A., Roggenbach, M. (eds.) FMFun 2019. CCIS, vol. 1301, pp. 208–217. Springer, Cham (2021). https://doi.org/10.1007/978-3-030-71374-4_11
14. García-Matos, M., Väänänen, J.: Abstract model theory as a framework for universal logic. In: Beziau, J.-Y. (ed.) Logica Universalis, pp. 19–33. Birkhäuser, Basel (2007)
15. Gasquet, O., Schwarzentruber, F., Strecker, M.: Panda: a proof assistant in natural deduction for all. A Gentzen style proof assistant for undergraduate students. In: Blackburn, P., van Ditmarsch, H., Manzano, M., Soler-Toscano, F. (eds.) TICTTL 2011. LNCS (LNAI), vol. 6680, pp. 85–92. Springer, Heidelberg (2011). https://doi.org/10.1007/978-3-642-21350-2_11

16. Gualà, L., Leucci, S., Natale, E.: Bejeweled, candy crush and other match-three games are (NP-)hard. In: IEEE CIG, pp. 1–8. IEEE (2014). https://doi.org/10.1109/CIG.2014.6932866

17. Hjelsvold, R., Nykvist, S. S., Lorås, M., Bahmani, A., Krokan, A.: Educators' experiences online: how COVID-19 encouraged pedagogical change in CS education. In: Norwegian Conference on Didactics in IT education (2020)

18. Hundhausen, C.D.: Toward effective algorithm visualization artifacts: designing for participation and negotiation in an undergraduate algorithms course. In: Karat, C., Lund, A. M. (eds.) CHI, pp. 54–55. ACM (1998). https://doi.org/10.1145/286498.286526

19. Hundhausen, C.D., Douglas, S.A., Stasko, J.T.: A meta-study of algorithm visualization effectiveness. J. Vis. Lang. Comput. **13**(3), 259–290 (2002). https://doi.org/10.1006/jvlc.2002.0237

20. Kahl, W.: Calculational relation-algebraic proofs in the teaching tool CalcCheck. J. Log. Algebraic Methods Program. **117**, 100581 (2020). https://doi.org/10.1016/j.jlamp.2020.100581

21. Landers, R.N.: Developing a theory of gamified learning: linking serious games and gamification of learning. Simul. Gaming **45**(6), 752–768 (2014). https://doi.org/10.1177/1046878114563660

22. Materzok, M.: Easyprove: a tool for teaching precise reasoning. In: 4th International Conference on Tools for Teaching Logic TTL. abs/1507.03675 (2015)

23. Myller, N., Bednarik, R., Sutinen, E., Ben-Ari, M. Extending the engagement taxonomy: software visualization and collaborative learning. ACM Trans. Comput. Educ. **9**(1), 7:1–7:27 (2009). https://doi.org/10.1145/1513593.1513600

24. Naps, T.L., et al.: Exploring the role of visualization and engagement in computer science education. ACM SIGCSE Bull. **35**(2), 131–152 (2003). https://doi.org/10.1145/782941.782998

25. Ölveczky, P.C.: Teaching formal methods for fun using maude. In: Cerone, A., Roggenbach, M. (eds.) FMFun 2019. CCIS, vol. 1301, pp. 58–91. Springer, Cham (2021). https://doi.org/10.1007/978-3-030-71374-4_3

26. Pierce, B.C., et al.: Software foundations (2021). https://softwarefoundations.cis.upenn.edu/lf-current/index.html

27. Prasetya, W., et al.: Having fun in learning formal specifications. In: ICSE-SEET, pp. 192–196 (2019). https://doi.org/10.1109/ICSE-SEET.2019.00028

28. Richardson, F.C., Suinn, R.M.: The mathematics anxiety rating scale: psychometric data. J. Couns. Psychol. **19**(6), 551–554 (1972). https://doi.org/10.1037/h0033456

29. Roggenbach, M., Cerone, A., Schlingloff, H., Schneider, G., Shaikh, S.A.: Formal Methods for Software Engineering. Springer (2022, to appear). https://link.springer.com/book/9783030387990

30. Smullyan, R.: The riddle of Scheherazade and other amazing puzzles, ancient & modern. New York (1997)

31. Spichkova, M., Zamansky, A.: Teaching of formal methods for software engineering. In: Maciaszek, L.A., Filipe, J. (eds.) ENASE, pp. 370–376. SciTePress (2016). https://doi.org/10.5220/0005928503700376

32. Wouters, P., van Nimwegen, C., van Oostendorp, H., van der Spek, E.D.: A meta-analysis of the cognitive and motivational effects of serious games. J. Educ. Psychol. **105**(2), 249–265 (2013). https://doi.org/10.1037/a0031311

33. Zach, R.: Boxes and diamonds: an open introduction to modal logic (2019)

34. Zhumagambetov, R.: Teaching formal methods in academia: a systematic literature review. In: Cerone, A., Roggenbach, M. (eds.) FMFun 2019. CCIS, vol. 1301, pp. 218–226. Springer, Cham (2021). https://doi.org/10.1007/978-3-030-71374-4_12

Increasing Student Self-Reliance and Engagement in Model-Checking Courses

Philipp Körner[1](✉)(iD) and Sebastian Krings[2](iD)

[1] Heinrich Heine University, Universitätsstraße 1, 40225 Düsseldorf, Germany
p.koerner@hhu.de
[2] Düsseldorf, Germany
sebastian@krin.gs

Abstract. Courses on formal methods focus on two aspects: teaching formalisms and exemplary applications as well as teaching techniques for implementing tools such as model checkers.

In this article, we discuss the second aspect and typical shortcomings of corresponding courses. As courses often focus on theoretical results, opportunities for working on real implementations are scarce. In consequence, students are easily overwhelmed with transfer tasks, e.g., when working on existing model checkers during theses or research projects.

We present several iterations of our course on model checking, including their goals, course execution as well as feedback from peers and students. Additionally, we discuss how the Covid-19 epidemic impacted our course format and how it was made more suitable for online teaching.

Finally, we use these insights to discuss the influence of formality on student engagement, and how to incorporate more practical aspects by introducing inquiry and research-based teaching.

Keywords: Education · Model checking · Inverted classroom · Experience report

1 Introduction

The development and improvement of model checkers [9] for the validation of hard- and software is an ongoing research topic in computer science [12]. Model checking research connects theoretical and practical aspects; new algorithms are often implemented inside well-known model checkers which have been in development for many years. Thus, they typically have large and often involved code bases posing an entry barrier for students.

This is seldom taken into account by university courses, which often remain on the theoretical level, not providing practical access for various reasons. The same used to hold true for our approaches to teaching model checking.

Different shortcomings of typical courses on formal methods and model checking have also been identified in a whitepaper published at FMFun [8]. Among

S. Krings—Independent Researcher.

J. F. Ferreira et al. (Eds.): FMTea 2021, LNCS 13122, pp. 60–74, 2021.
https://doi.org/10.1007/978-3-030-91550-6_5

other reasons such as limited exposure to formal methods and missing integration of FM courses with the rest of the curriculum, the whitepaper lists a number of shortcomings relevant to course design and execution:

– FM courses are usually very formal, especially when compared to the more hands on example-drive approach to teaching programming languages.
– FM courses often provide only limited practical experience.
– Initial interest is low, formal methods are perceived as inaccessible.
– Students are often unable to bridge the gap between theory and tools.
– A focus on technical details might distract students from key learning goals.

With our course on model checking, we try to overcome these restrictions by increasing student involvement and motivation. This is done by providing practical experience in developing, testing and using a model checker and abstracting away from one concrete formal method.

We would like to add, that project experiences, tool usage and working collaboratively have been identified as major areas in which students do not meet expectations from industry [21, 22, 25]. Those skills could be improved en passant when integrating practical (programming) aspects into an FM course.

The rest of the paper is structured as follows: First, we briefly present the context of our course on model checking in Sect. 2. Afterwards, we outline several couse iterations in Sect. 3, including their goals, course execution and feedback from peers and students. We focus especially on the latest iteration that shifted the course to a more suitable format for online-teaching, as necessary during the COVID-19 epidemic. After comparing the average grades for different iterations and trying to explain them in Sect. 4, we finally discuss the influence of formality on student engagement, and how to incorporate more practical aspects by introducing inquiry and research-based teaching in Sect. 5.

2   Context

At the Heinrich-Heine-University Düsseldorf, we teach two master's courses—each worth 5 ECTS—concerning formal methods: Firstly, "safety-critical systems", where modeling and validation of B [1] and Event-B [2] specifications is taught. This is done using both the animator and model checker PROB [17], as well as the Event-B IDE Rodin [7], Recently, PROB was also integrated into Jupyter notebooks [11], which can be used for teaching as well.

The second course, which is the main focus of this paper, is "Model Checking". This course roughly follows the first half of Principles of Model Checking (PoMC) [3]. It mainly deals with algorithms and techniques for implementing model checkers. After course completion, students should be able to unfold transition systems from a specification, classify different kinds of properties (safety, liveness, ω-regular), implement (fair) algorithms for invariant, safety and LTL model checking, and to formulate properties given in natural language in a suitable, mathematical notation. Depending on the lecturer, these two courses may overlap somewhat on the modeling aspect.

In contrast to safety-critical systems, the course on model checking has displayed the typical shortcomings discussed above: interest was low in general (i.e., a small number of enrollments), students attended the course in a strongly passive manner and grades were fluctuating.

3   Course Evolution

In this section, we present several iterations of our course. We highlight the ideas and goals, outline their execution and include available feedback.

Aside from informal ad-hoc feedback, we do not have data on the original lecture-based course. Official course evaluations during the semester require a minimum number of student answers that was not reached.

The first and last course iteration presented have additionally been supervised by the didactics department. This includes reviews by peer lecturers (who may teach in a different faculty) during the planning phase as well as during the execution of the course. The didactics department also requires resilient feedback in written or electronic form.

3.1   The Origins: a Classical Lecture-Based Course

Teaching & Course Execution. The baseline course is a "traditional", lecture-based setting with additional exercises. Exercise sheets were handed to the students on a weekly basis and discussed in the following week. Exercises were voluntary and often on a larger scale. A major point of criticism by students was that exercises could neither be solved nor discussed in time. Independent preparation was, nonetheless, expected.

The lecture itself was made up of the more formal and theoretical aspects of model checking: The most important definitions, examples, lemmas and theorems of PoMC [3] were put onto slides and proven. Algorithms were presented in pseudo-code, but students were not given the task to implement them.

No formalism was discussed in detail. Instead, snippets or short models of B, CSP or nanoPromela[1] were used to demonstrate how certain aspects of specifications may look like. Aside from some discussion on the state space explosion problem, neither practical experience, experiments nor implementation concerns were part of the course.

Grading. As all activity during the semester was voluntary, grades were given solely based on the performance in a 90 min written exam. The most common questions tested whether the student was able to extract a transition system from a specification, calculate the handshaking of two transition systems, understand Büchi automata, and check and (dis-)prove equivalences of LTL formulas or extract them from natural language.

---

[1] A subset of SPIN's [13] input language that is introduced in PoMC.

Reflection. We assume that the original course format is at least partially responsible for the relative unpopularity of the course: The lecture-based format, combined with a strong mathematical component and the resulting reputation to be (too) work-intensive, was certainly reducing motivation for attendance.

Following the FMFun whitepaper [8], focusing on the mathematical aspects of FM reduces motivation if done before students are able to see the benefits of fully formal approaches. Additionally, given the relative unpopularity of formal methods compared to hot topics such as data science and artificial intelligence [8], only few students enlisted at all.

During the course, students remained unengaged both during lectures and exercise sessions. Without hands-on experience, students cannot witness the benefits of such tooling themselves. Thus, students are unlikely to use or may even be deterred from exploiting model checking techniques after the course.

3.2 First Iteration: Introducing Research- and Inquiry-Based Learning

Idea and Goals. Several insights from trainings on teaching and learning made us realize that the model checking course should be reworked in order to increase student interest, engagement and to overcome the shortcomings discussed.

To improve, we decided to remodel our course. The overall goal was to move from classic lectures to active learning techniques for an improved hands-on experience. Furthermore, we noticed that students writing theses at our chair sometimes lacked the required knowledge about how research is performed and thus needed close supervision and initial training.

In order to motivate individual research and to enable students to train their skills, we moved from a lecture-based setting to inquiry-based learning. In particular, we intended for the course to follow the typical pattern of finding and asking research questions, collecting evidence or creating it through experiments, compare and discuss results as well as explain and publish differences discovered.

Teaching & Course Execution. The first iteration and its course execution has already been described in detail [16]. In the following, we thus only give a brief overview. In line with the intended learning outcomes, during the course, participants should:

- Acquire the theoretical foundations of model checking by identifying and analyzing common software errors.
- Align these foundations with the body of knowledge.
- Design and implement a novel model checker as independently as possible.

To reach the goals, we started with an introductory example. Together with the students, we brainstormed numerous hazards the control software of a lift could suffer from. Following, we used an example-driven approach to introduce the B language. Since students were mainly supposed to learn about model checking algorithms and implement a model checker, technical details on B's semantics were not relevant.

Afterwards, students were asked to describe the behavior of the control software in (their impression of) B. Once consensus on the software specification was reached, students were asked to (independently) research verification algorithms and to implement them collaboratively. While doing so, students realized what is needed to turn the hazard collection into a checkable specification: invariants, (temporal) logic, etc.

During their research, we had different sessions where students could discuss their findings and ideas with us and where we provided further input and clarification. Additionally, these sessions where used to ensure student research was going into the intended direction.

Grading. When planning courses and exams, the learning outcomes, teaching methods and assessment should be constructively aligned [5]: Roughly summarized, student's activities during the course should reflect the intended learning outcomes. Simultaneously, the exam should be similar to those activities as well. Otherwise, students will learn what they think will be needed to pass the exam rather than what the course is supposed to teach them.

Thus, a more practically oriented course needs an appropriate grading method: We decided to use a combination of grading participation in the research and programming projects and a classical exam for the more theoretical parts.

Feedback

Peer Review. Two separate sessions were monitored and reviewed:

- An R & D session in which the students drove the prototypical model-checker forward by discussion algorithmic approaches and further implementation.
- A session meant as a synchronization point between two groups working on infinite-state approaches to model checking or time-based reasoning.

For the R & D session, the overall concept of individual inquiry and research by students combined with discussing existing approaches once they were "discovered" was seen as innovative, effective and appropriate for the goals we set. However, some weaknesses were spotted as well:

- Sometimes, we failed to ensure that all participants had understood a topic well enough to participate further. This led to diverging groups, in which novel algorithms were developed by those still able to follow the train of thought. Simultaneously, some students remained on their own and did not take part in the discussions until the group met again. For future sessions, we decided to discuss, document and visualize new ideas more thoroughly.
- For some research ideas our students developed, next steps remained unclear and nobody was assigned to drive them forward. Essentially, lacks in overall project management accounted for ideas getting lost and rediscovered later on. We improved the project management tools used (Kanban boards with dedicated assignees) to overcome this issue.

The synchronization session was meant to update different focus groups with the results of the other students. Furthermore, intermediate presentations should help account for the appropriate reflection upon the executed research tasks. As pointed out in [6,23], this helps to avoid focusing to heavily on execution without evaluating results and lessons learned. These sessions increase student's opportunity for self-assessment and revision, one of the principles of successful inquiry-based courses stated by Barron et al. [4].

As stated by the lecture reviewers, student presentations were sluggish, most likely caused by their inappropriate preparation. In retrospect, we identified our imprecise and not explicitly given expectations as the most likely reason. We suspect that students did not get the overall concept of the presentation & synchronization lessons. Without realizing what we aimed at, they were unable to perform appropriately.

Again, we adapted following sessions by supplying a coarse outline when asking for presentations: presentation of technique used, blockers and intended solutions used and open issues that could then be solved by the whole group.

To increase commitment, presentations had to be discussed with the lecturers beforehand in order to ensure quality requirements were met. Overall, following sessions were able to distribute individual knowledge to the other participants as intended.

Student Feedback. For the inquiry-based course, we performed several intermediate online evaluations. We asked students for their workload, motivation and an individual estimate of their learning outcome. Each was to be rated on a scale from 1 (lowest) to 5 (highest). Furthermore, we asked students to give a reasoning for their ratings using free-text answers.

The overall feedback was very positive and encouraging. In particular, the switch from a lecture-based to an inquiry and research-based design was successful:

- The course was described as interesting,
- Student evaluated their individual learning outcome as high (average of 4.375), mostly attributed to the increased self-reliance.
- The course was described as very work intensive, with an average of 3.75. However, the overall hours dedicated to the course by students was in alignment with the ECTS awarded. Thus, we deem the workload appropriate.
- Even though workload was high and the course was described as quite demanding, overall motivation was rated with an average of 4.625.

The most dominant point of criticism was the volatile speed of progression. While it is certainly impossible to guarantee a progression speed while allowing individual research, the overall progress had to be ensured better and in a way that was obvious for the students. This criticism drove some changes in the later course iterations.

We were pleasantly surprised by the very highly rated motivation. However, individual motivation is hard to compare to other courses judging by the self-assessment alone.

Thus, we tried to empirically evaluate student motivation using the activity and commit data available for the repository[2] used to develop the model checker. To summarize [16]:

- Most changes were made during the sessions in presence.
- Consistent activity throughout weekdays and working hours.
- Higher activity before lectures, maybe due to students revising the material.
- Occasionally, we see a student working all night. In a second evaluation, students stated that this was not caused by an overboarding work load, but to individual interest, motivation and time management.

Another indication for high motivation due to our research-based approach is that three of the participants worked with us towards a publication of their research results. This was an optional offering, not linked to the course, its grading or rewarded with credit points. Still, students put in further effort and managed to publish a paper [20].

Reflection

Scalability and Repeatability. We do not deny that a group of students is able to write an interesting, somewhat sophisticated and useful program during a semester. However, due to the size and requirements of the programming project, different skills and programming experiences were needed for success. Yet, we cannot reasonably expect those skills to be available in following iterations of the course. Additionally, performing a joined programming project does not scale arbitrarily. This approach would not work with semesters where too few or too many students attend. Of course, one could create several group competing in, e.g., performance. Yet again, different skills should be present in each group and one would have to deal with assigning students to groups accordingly.

Grading. Grading students based on their involvement in a programming project is hard and often not as objective as desired. Objective metrics, such as the number of commits or lines of code give no insight into the students' knowledge and understanding.

Knowledge Propagation. Students tend to acquire more knowledge about their own area of focus. Since the research work was split into different topics, knowledge often did not propagate equally. In our experience, this holds true for both seminars and programming projects.

Too Rich Formalism. The B language is very expressive. While this allows for concise and precise specifications, evaluating state transitions is complicated and needs constraint solving algorithms, e.g., to compute parameters. Thus, even the subset of B given to the students required considerable work on implementing a language interpreter, effectively diverting resources from the actual model checking algorithms. Yet, this iteration raised a very important question that we shall return to later on: *"How formal should formal methods be taught?"*

---

[2] https://github.com/bmoth-mc/bmoth.

3.3   Second Iteration: Lessons Learned: Mixing Lecture and Practical Exercises

Goals and Ideas. Due to the concerns mentioned above, we decided not to repeat the course without further modification. Instead, we tried to combine the best of both worlds by teaching theoretical aspects using lectures while also making students work on individual, smaller-scale programming projects.

Teaching & Course Execution. With this iteration, we started from the initial set of lectures again. We cut back on long proofs, reduced the build up to important theorems and provided smaller exercises that still kept the main idea.

Instead, students were supposed to individually implement a model checker with LTL capabilities. This time, the formalism was kept intentionally simple (i.e., petri nets), so a naive reachability tool can be implemented in a few hours.

We also provided parsers for the models and LTL formulas, as well as a transformation of formulas in positive normal form to generalized non-deterministic Büchi automata. Learning from the overhead of implementing a language interpreter, this allowed students to focus on the model checking aspects.

This programming project is sufficiently small that it could be done in an appropriate amount of time, yet also forces students to internalize the required steps for LTL model checking.

Grading. Additionally to the summative exam at the end of the semester, we kept a formative part of the grade: A reduced version of the model checker project, that can be used for reachability analysis, deadlock and invariant checking, was mandatory to take an exam. Full LTL capabilities of the resulting tool made up 20% of the overall grade.

Student Feedback. Two points stand out in the evaluation: firstly, all students still attending the course rate it excellent in structure, materials, lecturer and overall impression. Furthermore, students rate their subjective learning success as excellent to good.

Additionally, the programming project was highlighted positively in an open question. However, compared to other courses, the initial interest is rated rather low: on a scale of 1 (best) to 5 (worst), the median 3 and arithmetic mean 2.4.

Reflection. Overall, we think this course is an appropriate compromise. It ensures that students follow the correct path, yet enforces hands-on experience. However, grading proved to be difficult.

Grading. When reviewing and testing the students' code it is usually all or nothing. Grading programming projects without clearly communicated criteria is not feasible. One has to decide, whether only to grade functionality and correctness, or to add criteria such as code style, performance, etc., and how to grade those.

3.4   Third Iteration: Improved Teaching Methods and Online Teaching

Goals and Ideas. In 2020, due to the COVID-19 pandemic, we had to make a quick leap to online teaching with very limited preparation. Additionally, the course was held in 13 rather than 15 weeks, as the semester was shortened because of the pandemic. This did not allow us to continue the mode of the prior iterations and forced us to try something new. In particular, these conditions did not allow us to include programming work without abandoning important content.

The situation suggested an inverted classroom (see, e.g., [26]): Instead of giving a lecture and having the students prepare exercises at home, we met for weekly exercise sessions with reading tasks to be done individually. With online versions of standard books on model checking [3,9] and lecture recordings by the RWTH Aachen University[3], this was possible without long preparation.

Execution. In this iteration, the course followed the first five chapters of PoMC with a final glimpse at timed automata. Note that this involves a large share of mathematical notation, proofs, etc. Course organization and philosophy were heavily inspired by Keller's Personalized System of Instruction [14]:

On-line Sessions. One goal was to limit the time spent interacting online and invest it into self-study instead. In particular, no lecture was held. This decision comes with the idea that the—mostly mathematical—load during a regular lecture is far too high for a student to actually follow and, thus, participate. In the past, certain proofs (e.g., correctness of the nested depth-first search) were guaranteed to outpace the majority of students. Instead, a typical session was structured as follows:

1. Opening and mood barometer: as part of the opening, it was important to us to poll the current mood of the students, and, to build trust, give our own. Usually, it was used to gain insights on key questions (e.g., the students' perception of the new course iteration, their current situation during the pandemic, their happiness with online courses overall). For this, a slide with a 4×4 grid of emojis was prepared that students were able to point to and draw at. Additionally, everyone could comment via voice or chat. This method allowed some initial activation and personal interaction, which—in our case—led to a good course atmosphere.
2. Material review and electronic voting system (EVS): Miller and Cutts described benefits of an EVS in a formal methods course [18]: In particular, they used it to ensure that students read the material and understood it, and students became more confident in their knowledge and were more willing to answer questions. Thus, in the next part, we prepared some single-choice questions on the material for revision and as a light means of testing

---

[3] https://www.youtube.com/playlist?list=PLnbFC0ntxiqdpoWwMKCVh6BRwBeP HaqQx.

understanding. Occasionally, questions were designed to trap students with fallacies that we have observed in the past[4]. Naturally, these were discussed more in-depth afterwards.

3. Live exercises: For the majority of the session, students were given the opportunity to ask questions on the material and to choose the exercises they want to solve (e.g., taken form the large number of questions in [3]). For the key concepts, we prepared questions that showed on a minimal example how a technique works. Proofs and answers were written co-operatively on a shared whiteboard, with the lecturer only adding notes, guiding the students if stuck or correcting errors.

4. Outlook and intuition: Again, the idea is that the most important benefit of a lecture is that students develop an intuition, yet have to work out details at home at their own pace. During the last minutes, we tried to give an intuition on the material that should be prepared for the next session. We only sketched connections to prior material and the basic idea, without formal definitions, proofs, etc. Instead, we raised key questions the students should find the answer to.

Learning Units. The course was structured into units that were made available for the students entirely at the beginning of the semester. Each unit contained references to learning material (i.e., relevant sections of books, lecture recordings and scientific articles), a list of expected learning outcomes, a brief enumeration of the most important concepts of the unit, and a collection of exercises.

Learn at Your Own Pace. Students were given the opportunity to choose their own pacing which they deem suitable for their learning style. This includes both the individual speed (which—in a traditional setting—is dictated by the speed of the lecturer) and the time during the semester they learn (each week, or starting a few days before the exam).

To motivate continuous work and ensure students prepare for the online sessions, they were allowed to hand in a learning diary before the session. This learning diary may include anything related to the learning unit (though not simply copies of textbooks), e.g., notes on definitions or solutions to exercises they solved themselves, and was allowed as individual resource during the exam. A short version limited to two pages was allowed to be handed in up to a week afterwards, in case personal circumstances (sickness, etc.) rendered it impossible for some students to complete it in time. All learning diaries were inspected by a teaching assistant and errors were marked.

Feedback

Peer Review. For the online setting, peers and the didactics department were satisfied with the methodology. They also were able to identify issues related to eLearning. Their highlights include the following:

---

[4] Aiming at finding questions and distractors to eventually be used for peer instruction [10].

- The mood barometer is a nice method for activation, with the students and lecturer even revealing personal insights.
- The use of revision and EVS is good. Yet, during discussion of follow-up questions students often are unsure whether to use their microphone, chat or wait for voting options.
- The methodology of assigning the task of creating their own summaries (e.g., what are the steps required for LTL model checking) and take home-messages (e.g., how do counterexamples to different kinds of properties look like), to students is important to deepen their understanding.
- The students are very engaged and take initiative during the sessions in creating and discussing solutions.
- Nonetheless, student webcams remain deactivated.

Student Feedback. For the online course during 2020, we polled the students after the course. A four-point scale of strong/weak agree/disagree with an option of not applicable was used. Students unanimously (strongly, unless stated otherwise) agreed on the following:

- The structure of the course is excellent,
- the inverted classroom setting was worthwhile,
- the online session was very useful (one weak agreement),
- the PoMC book is very understandable (one weak agreement),
- the lecture recordings of Prof. Katoen are very understandable,
- explicitly-stated learning outcomes helped their self-study,
- overall, they are very satisfied with the course.

On the following two statements, one student weakly disagreed while all others agreed: "An inverted classroom would be worthwhile in an off-line setting", and "The outlook part of the session was helpful".

In an open question, a student highlighted that they felt that nobody was left behind in case there were questions of uncertainties and that everything was discussed until everyone understood the matter.

Reflection

On Establishing a Testing Culture. Due to a small course size, the learning diaries and a good atmosphere during the online sessions, it was possible to ensure that all learning outcomes were met before the exam, both for the students and the lecturer.

In retrospect, we would recommend online testing of students, where students have to achieve very high marks for every unit, yet may attempt a test as often as necessary. In another course, it has proven valuable to discuss every test briefly (i.e., 5–10 min) with the individual student, probing for understanding of the matter and correcting minor mistakes. Teaching assistants can share this load with a lecturer or even do the work entirely.

Table 1. Average Grades (in Parentheses: Adjusted Data Without Failing Students/Without Formative Grades)

Year	# enrollments	# exams	∅ grade	course type	exam modus
2014	11	2	1.85	lecture-based	written
2015	12	(4) 5	(1.98) 2.58	lecture-based	oral
2016	13	7	1.71	lecture-based	written
2017	11	6	(1.43) 1.28	inquiry-based	written + formative
2018	18	5	1.88	lecture + programming	written
2019	10	5	1.58	lecture + programming	written
2020	16	5	1.54	online + inverted classroom	written

Exemplary Solutions. An opinionated topic is whether, when and how solutions to exercises should be made available to students (e.g., [19]). While we cannot give an ultimate answer to these questions, we noticed that, often, exercises requiring proofs (e.g., classification of properties as safety or liveness properties) contained errors—even though a solid mathematical education is required to attend our course. To counteract, we think that in these cases an annotated and correct solutions should be made available to the students, even if it is to simply ensure that they have seen a correct solution and what fallacies need to be avoided.

No Hands-on Activity. Even though no programming activity was mandatory during the semester, students were aware that they might face programming tasks in the exam as stated in the expected learning outcomes. In fact, when provided with a small interface, all students were able to implement a variant of the model checker, that was required for admission the years before, without errors.

4 Comparison of Grades

One of the measurements—and sometimes the only—of learning success are the grades given at the end of the semester. In this section, we give an overview of the grades throughout the years and discuss how they can be compared.

Grades are given on a scale from 1 (excellent) to 5 (fail) in steps of 0.3. An overview of the last years is given in Table 1. Note that in 2015, a failing student may distort the data, and the adjusted value is given as well. In 2017, when including the formative part of the course, i.e., implementation work and participation, the average grade improves as well. The value of the exam alone is additionally given in parentheses.

In general, it is hard to draw reliable conclusions from relatively small sample sizes of five to seven exams per year. However, one can identify certain trends that align with different teaching methodology:

When considering the data, one can see that students attending the inquiry-based course in 2017 were more successful than those attending the purely lecture-based courses since 2014. Especially with the formative part, the average grades improve significantly. One explanation might be that the practical research experience deepened the students' understanding of the course matter much better than attending lectures. On the other hand, the student feedback also discloses a much higher workload than their other courses.

Students attending the second iteration in 2018 were on par with the lecture-based courses and improved in 2019. A possible reason for this might be the additional experience gained with this teaching style and, thus, better supervision of student activities. In the online course from 2020, students performed slightly better, and the amount of excellent grades increased as well.

One interesting outlier is the course from 2015, where oral exams were held rather than written ones. An explanation for the significant lower grades might be that students were not used to talk about course matter, and thus achieved lower grades in that setting.

5   Conclusions

In this paper, we presented several iterations of our course on model checking, describing a shift from lecture-based teaching to more self-responsible learning, increasing participation and practical experiences. With these experiences, we will return to the issues raised in Sect. 1 and present our conclusions.

How Formal Should Formal Methods be Taught? We think that student engagement in lectures heavily benefits from being more informal, i.e., avoiding strictly mathematical discussions and long proofs. All hands-on exercises, may it be programming tasks or exercise questions that engage students in discussion, are preferable. The formal aspects of formal methods can be taught as part of reading tasks instead, as we have done in the online course. Additionally, we argue that the formal parts of formal methods should be taught only after the benefits of using formal methods are understood by the students.

To What Extent do Students Benefit from Practical Experience? As reported, student feedback on practical projects has always been positive.

Practical experience in implementing presented algorithms provides a second gateway to the desired learning outcomes[5]. We suspect that after implementation students have more in-depth knowledge, have realized certain edge cases and are more likely to remember technical details.

As an alternative to implementing model checkers from scratch, examining existing tools such as PROB could be considered. This however comes with an entry barrier. For instance, PROB is implemented in Prolog which students usually do not know well. Furthermore, the code base is large and code snippets might need a lot of context to be understood.

---

[5] Keep in mind that modelling systems and applying model checkers to them is taught in a seperate course.

Also, modern model checkers rely on many other techniques students did not implement, e.g., BDDs or partial order reduction. In summary, existing implementations are often more confusing than helpful and clean room implementations are to be preferred.

Overall, the benefit of having students implement a model checker may be lower than we hoped for when considering potential for student theses. Nonetheless, it seems to be a valuable addition to a course on model checking.

How to Increase Student Interest? How to Improve Student Perception of Formal Methods? This question cannot be answered conclusively from our experience. Simply put, it is too late to reach students once they decide to choose another course over a formal methods one. One possibility is to tighter integrated formal methods with other courses in the computer science curriculum (as suggested by [8]). An alternative could be to advertise FM courses with innovative and fun teaching concepts.

However, once they reach the classroom, we believe that with properly designed courses we can convince students that formal methods are more than boring mathematics, and tools are not black magic.

There are many suggestions for courses focusing on modeling and proof, e.g., using games [15] or puzzles [24] as examples. Judging from student feedback, we think that programming tasks or letting students find answers to certain questions themselves are an engaging and, ultimately, fun approach.

Inverted Classrooms and Online Sessions. We argue that, especially for online teaching, an inverted classroom is a viable alternative—especially, when a course follows a more formal approach. In a lecture-based approach, there is rarely time to comprehend new mathematical formulas and proof during lectures. Yet, even in a less formal format, students can heavily benefit from acquiring theoretical knowledge at home and use the time of synchronous sessions for discussions instead.

Acknowledgement. The authors would like to thank their peer lecturers Jens Bendisposto, Natalie Böddicker, Janine Golov, Ann-Christin Uhl and Susanne Wilhelm for reviewing their courses. They also thank Joshua Schmidt for his input, fruitful discussions and course execution in 2019.

References

1. Abrial, J.R.: The B-Book: Assigning Programs to Meanings (1996)
2. Abrial, J.R.: Modeling in Event-B: System and Software Engineering. Cambridge University Press, Cambridge (2010)
3. Baier, C., Katoen, J.P.: Principles of Model Checking. MIT press, Cambridge (2008)
4. Barron, B.J., et al.: Doing with understanding: lessons from research on problem- and project-based learning. J. Learn. Sci. **7**(3–4), 271–311 (1998)
5. Biggs, J.: Enhancing teaching through constructive alignment. High. Educ. **32**(3), 347–364 (1996)

6. Blumenfeld, P.C., Soloway, E., Marx, R.W., Krajcik, J.S., Guzdial, M., Palincsar, A.: Motivating project-based learning: sustaining the doing. Support. Learn. Educ. Psychol. **26**(3–4), 369–398 (1991)
7. Butler, M.J., Hallerstede, S.: The Rodin Formal Modelling Tool 1. BCS-FACS Christmas Meeting (2007)
8. Cerone, A., et al.: Rooting formal methods within higher education curricula for computer science and software engineering - a white paper. CCIS, vol. 1301 (2021). https://arxiv.org/abs/2010.05708. Springer
9. Clarke, Jr., E.M., Grumberg, O., Peled, D.A.: Model Checking. MIT Press, Cambridge (1999)
10. Crouch, C.H., Mazur, E.: Peer instruction: ten years of experience and results. Am. J. Phys. **69**(9), 970–977 (2001)
11. Geleßus, D., Leuschel, M.: ProB and jupyter for logic, set theory, theoretical computer science and formal methods. In: Raschke, A., Méry, D., Houdek, F. (eds.) ABZ 2020. LNCS, vol. 12071, pp. 248–254. Springer, Cham (2020). https://doi.org/10.1007/978-3-030-48077-6_19
12. Grumberg, O., Veith, H. (eds.): 25 Years of Model Checking: History, Achievements, Perspectives, LNCS, vol. 5000. Springer, Heidelberg (2008)
13. Holzmann, G.J.: The model checker spin. IEEE Trans. Softw. Eng. **23**(5), 279–295 (1997)
14. Keller, F.S.: Good-bye, teacher.... J. Appl. Behav. Anal. **1**(1), 79 (1968)
15. Krings, S., Körner, P.: Prototyping games using formal methods. In: Cerone, A., Roggenbach, M. (eds.) FMFun 2019. CCIS, vol. 1301, pp. 124–142. Springer, Cham (2021). https://doi.org/10.1007/978-3-030-71374-4_6
16. Krings, S., Körner, P., Schmidt, J.: Experience report on an inquiry-based course on model checking. In: Proceedings SEUH 2019, vol. 2358, pp. 87–98. CEUR (2019)
17. Leuschel, M., Butler, M.: ProB: an automated analysis toolset for the B method. STTT **10**(2), 185–203 (2008)
18. Miller, A., Cutts, Q.: The use of an electronic voting system in a formal methods course. In: Proceedings FM-Ed 2006, pp. 3–8 (2006)
19. Nygren, H., Leinonen, J., Hellas, A.: Non-restricted access to model solutions: a good idea? In: Proceedings ITiCSE 2019, pp. 44–50. ACM (2019)
20. Petrasch, J., Oepen, J.H., Krings, S., Gericke, M.: Writing a Model Checker in 80 Days: Reusable Libraries and Custom Implementation. ECEASST (2018)
21. Radermacher, A., Walia, G.: Gaps between industry expectations and the abilities of graduates. In: Proceeding SIGCSE 2013, pp. 525–530. ACM (2013)
22. Radermacher, A., Walia, G., Knudson, D.: Investigating the skill gap between graduating students and industry expectations. In: Proceedings ICSE 2014, pp. 291–300. ACM (2014)
23. Schauble, L., Glaser, R., Duschl, R.A., Schulze, S., John, J.: Students' understanding of the objectives and procedures of experimentation in the science classroom. J. Learn. Sci. **4**(2), 131–166 (1995)
24. Schlingloff, B.-H.: Teaching model checking via games and puzzles. In: Cerone, A., Roggenbach, M. (eds.) FMFun 2019. CCIS, vol. 1301, pp. 143–158. Springer, Cham (2021). https://doi.org/10.1007/978-3-030-71374-4_7
25. Tafliovich, A., Petersen, A., Campbell, J.: On the evaluation of student team software development projects. In: Proceedings SIGCSE 2015, pp. 494–499. ACM (2015)
26. Talbert, R.: Inverted classroom. Colleagues **9**(1), 7 (2012)

Teaching Formal Methods to Software Engineers through Collaborative Learning (Short Paper)

Livia Lestingi(✉) (iD)

Dipartimento di Elettronica, Informazione e Bioingegneria, Politecnico di Milano,
Milan 20133, Italy
`livia.lestingi@polimi.it`

Abstract. It is common knowledge among researchers in the field that teaching formal methods can prove a challenging task. This paper reports on the approach adopted for a Master's Degree course at Politecnico di Milano, Italy, as an attempt to reverse this trend by introducing collaborative learning activities. Students put concepts learned during theoretical lectures into practice through a hands-on group assignment. Each group develops the formal model of a Cyber-Physical System through the Uppaal tool, starting from a set of requirements provided by the instructor. After delivering the assignment, we invite students to fill an evaluation survey whose results suggest a very high satisfaction level towards the hybrid theoretical-practical approach of the course.

Keywords: Formal methods teaching · Collaborative learning · Software engineering education · Postgraduate education

1 Introduction

Formal methods is not what students in Computer Science are most passionate about. Over the years, several experts in the field have tried to identify the root of the problem, and the effort required to grasp the **mathematical notation** is the most commonly mentioned issue. Especially for Software Engineering education, there seems to be an ever-growing gap between the practical approach of software development and the theoretical approach of formal methods research [16]. The negative perception of mathematics is so extensive that *"mathematical anxiety"* is now a customary expression. This perspective is contradictive considering that Engineering students deal with mathematics daily and that other branches of Computer Science, such as Machine Learning, are not inferior to Formal Methods in terms of mathematical complexity but widely more popular.

Over the years, several different teaching strategies have been proposed as possible solutions to this issue. Mandrioli [15] suggests adopting an *incremental* approach and increasing the level of *user-friendliness* (for example, by favoring *state-based* notations over formulae) without forsaking the rigor of mathematical

J. F. Ferreira et al. (Eds.): FMTea 2021, LNCS 13122, pp. 75–83, 2021.
https://doi.org/10.1007/978-3-030-91550-6_6

modeling. Liu et al. [14] suggest gradually introducing students to the most complex concepts, increasing the number of exercise sessions, and helping students understand the power and effectiveness of formal techniques through short and simple examples from daily life (also previously suggested by Gibson and Mery [7]). Others propose to increase tool support during teaching activities. The following tools are notable examples: the KeY-Hoare tool [9] for teaching Hoare logic; a toolset developed by Korečko et al. [11] for teaching formal aspects of software development based on Petri nets and B-Method; a tool by Spichkova et al. [18] for model-based testing that also accounts for possible human mistakes.

This paper reports on the approach adopted for the *Formal Methods for Concurrent and Real-Time Systems*[1] course for Computer Science Master's Degree students at Politecnico di Milano. The teaching approach features the innovative element of **Collaborative Learning** [12], which allows students to work together in small groups towards a practical goal exploiting the concepts learned during the theoretical lectures. At the end of the course, we have invited students to fill an evaluation questionnaire to assess their satisfaction level. The collected results paint a very positive picture: the vast majority of students who participated in the survey reported increased **confidence** in course topics and a genuine interest in the activities undertaken while working on the project.

The paper is structured as follows: Sect. 2 presents the context of the course, i.e., the Computer Science program and the Software Engineering curriculum; Sect. 3 introduces the innovation of Collaborative Learning; Sect. 4 illustrates the educational goals, requirements and theme of the group assignment; Sect. 5 presents the results of the evaluation survey; Sect. 6 concludes.

2   Course Context and Structure

The Computer Science and Engineering M.Sc. program at Politecnico di Milano allows students to build their curriculum flexibly. Courses are grouped into ten **tracks** that students can choose from to elect their specialization. Tracks cover a wide range of Computer Science branches and are updated yearly to keep up with the latest technological trends. Formal Methods for Concurrent and Real-Time Systems is part of the *Software Engineering for Complex Systems* track. The track's goal is to train future engineers to tackle issues related to the development and deployment of **complex software systems**.

The course is structured to teach students how to exploit formal methods throughout the **software development** process. The relevance of this practice has already been acknowledged over the years [10] and is now gaining more popularity as the demand for dependable software increases. A recent survey by Gleirscher and Marmsoler [8] highlights a non-negligible usage of Formal Methods in some areas, such as transportation and critical infrastructures. Promoting education on these techniques might help break the vicious cycle formed over the years [19]. Specifically, FM cannot spread in industry if employees (i.e., former students) do not possess sufficient knowledge on the topic, while students are

---

[1] Full information about the course can be found at: https://bit.ly/3gLXOdR.

not motivated to study it if (among the other reasons) expertise on FM is not required to work in industry. As a matter of fact, although more than 400 students enrol in the CS M.Sc. program every year, only about 40 of them chooses the FM course for their curriculum (specifically, 39 for A.Y. 2020/2021).

The selection of topics for the course (fully reported by Askarpour and Bersani [3]) and the adopted teaching approach are a tentative compromise between the two conflicting tendencies in FM teaching: deep focus on **theoretical background** and mathematical formalism versus the **learning by doing** approach. Concerns that Computer Science education is exceedingly distancing itself from abstract theoretical concepts started to emerge twenty years ago [20]. Over the years, this dichotomy has sparked a debate on whether this ultimately results in less-prepared computer scientists [15] or boosts their chances to solve real-life problems successfully [4]. In the following, we present the strategy adopted to tackle this challenging issue.

3 Introducing Collaborative Learning

The initial assumption is that avoiding the theoretical side of FM topics is not a viable option. Indeed, only teaching students how to use verification tools without proper knowledge of the underlying formalisms defeats the purpose of an academic formal methods course. On the other hand, the lack of *confidence* caused by the often elaborate mathematical notation requires attention and explaining theory through small examples only partially solves the problem [14].

The adopted strategy features two alternative ways for students to pass the course. They can either do an **oral examination** on course topics, which counts for 100% of the final grade or: 1) select and **present** a **FM tool**[2] in front of the classroom, counting for 60% of the final grade; 2) work on the **group assignment**, counting for 40% of the final grade. Both presentation and project have to be carried out in groups of 2–4 people. Students have about two months to work on the project before the final deadline. As of this year, 75% of the classroom has chosen the second alternative (tool presentation and group assignment).

The innovative measure is the introduction of **Collaborative Learning** (CL). CL is *"an educational approach to teaching and learning that involves groups of learners working together to solve a problem, complete a task, or create a product."*[12] Students work on the same task which is entirely carried out using only one tool making it impossible to delegate fully independent subtasks to single group members; thus, students have to rely on one another to achieve the goal promoting **interdependence**. Students cannot complete the task autonomously, but they have to **interact** and challenge each other's ideas. The course targets students aged (on average) 23-24 who have completed at least three years of academic education; thus, they naturally display a good level of individual **accountability**. Group discussion and collaborative thinking help students develop skills such as conflict management and leadership. Finally,

---

[2] Eligible tools include: JBMC, CBMC, Prism, TLA+, COSMOS, SPIN, NuSMV.

groups are encouraged to monitor their progress with respect to the delivery deadline and periodically contact the instructor to receive feedback.

At the time of project assignment, Italy was not in full lockdown due to the COVID-19 pandemic. Although we do not possess accurate data due to privacy concerns, it is safe to estimate–based on how many people physically attended lectures[3]–that the vast majority of them were not residing permanently in Milan during project development. Nevertheless, the project outcome is entirely in software form and the university provides all students with the tools necessary to communicate remotely. Therefore, the collaborative learning strategy has not been affected by COVID-19 limitations at its core.

4   Group Assignment: Goals and Structure

This section reports on the group assignment's educational goals and how it is structured to meet these requirements. Afterward, we report the specific theme and model requirements for this year.

4.1   Educational Goals

The group assignment fits in with the Software Engineering profile of the students attending this course since its educational **goals** are:

G1: developing the **modeling** skill, i.e., how to translate informal requirements (expressed in natural language) into the formalism of choice;

G2: amplifying **critical thinking**, in terms of analytical experimental data evaluation to gain insights into the system performance;

G3: improving the capability of **expressing** oneself clearly and convincingly in written form and using accurate scientific language.

The centrality of managing models for Computer Science (goal **G1**) is widely acknowledged [5]. Besides requirement abstraction, students must perceive that a model can never perfectly match reality. Therefore, they must quickly learn how to balance complexity. System's behavior should be verified or simulated but purely reporting the results without analyzing them defeats the purpose of an engineering degree [6]. Students should proactively experiment with different system configurations and assess how this impacts the performance (goal **G2**). Finally, a study has revealed how non-technical skills, such as *self-expression* (goal **G3**), are highly requested by job applications, but they are also one of the most common gaps in a fresh engineering graduate's skillset [17].

Each **deliverable** required by the assignment fulfills one of the set-out goals:

D1: the developed **formal model** meeting the initial set of requirements;

D2: **verification results** and multiple (≥ 2) **model configurations**;

---

[3] An internal survey shows that physical attendance rate at its peak was only 25%, whereas 75% of the students attended remotely.

D3: a **written report** describing model, experimental results, design choices, and the reasoning that led the team to choose one alternative over the others.

In the following, the specific project theme is presented with technical aspects such as the selected formalism and verification tool.

4.2   Project Content: Model-Checking for Warehouse Robotics

This year's (A.Y. 2020/2021) project theme is **warehouse robotics** management. Automated warehouses significantly spread over the last few years thanks to the introduction of mobile robots. These wheeled platforms can take charge · of several tasks, most importantly *picking and delivering* operations. Items are stored in racks (i.e., *pods*) that robots can lift and transport to a *delivery point*. Human operators are usually in charge of manual edge tasks in this setting, such as picking the specific item from the pod. Students are required to develop a formal model (deliverable **D1**) of the following entities:[4] a) the warehouse **layout**: b) the **robots**; c) the **tasks**; d) the **human** operator. Specifications intentionally leave room for interpretation. For example, each team is free to choose whether the layout should be a standalone automaton or hard-coded into the model (i.e., as a two-dimensional array). The goal is to push each team to make **design choices** while drafting their model and provide reasonable justifications.

For the past ten years, the project focused on temporal logic [3], while, as of the last two editions of the course, the formalism of choice is **Timed Automata** (TA) [2]. The system under analysis dictates this choice since its behavior mainly hinges on timely synchronization among the different elements. Several features also naturally lend themselves to be expressed as *clock constraints*. For example, robots move every K time units, a new task spawns every T time units, and the operator takes time H to pick the item, where K, T, H are constant parameters.

Subsequently, students have to verify through **model-checking** a critical property (deliverable **D2**). The mandatory property is: *"it never happens that the number of tasks in queue exceeds the maximum queue size."* This property subsumes that the chosen system configuration (e.g., number of robots, robot speed, tasks spawn rate, etc.) allows robots to complete tasks quickly enough to avoid *task overflow*. The property must be expressed in TCTL (Timed Computation Tree Logic). For illustrative purposes, a possible formulation is shown in Eq. 1, where parameter MAX_T corresponds to the queue size and the number of tasks currently stored in the queue is captured by variable n_{tasks}.

$$\forall \, \square \, (n_{\text{tasks}} \leq \text{MAX\_T}) \tag{1}$$

Both the modeling and verification tasks of the assignment must be entirely carried out through the **Uppaal** tool [13]. As mentioned in Sect. 2, the course program includes a whole session dedicated to a hands-on demo of the tool.

---

[4] The full set of requirements is available at: https://bit.ly/3mEJbgc.

Starting this year, we have included the option to add **stochastic** features to the developed model that count as extra points in the final evaluation. These optional model features capture the **uncertainty** (refined by probability distributions) of the system's behavior. The introduction of probability distributions makes the automaton network no longer eligible for exhaustive model-checking but fit for **Statistical Model Checking** (SMC) [1]. To this end, students attend two additional lectures on the fundamentals of SMC and the Uppaal SMC extension. If they choose to pursue the stochastic path, they must verify through SMC the *probability* of property in Eq. 1 holding within a time-bound τ, whose formulation in PCTL logic is given in Eq. 2.

$$\mathcal{P}_{\leq \tau}(\square \; n_{\text{tasks}} \leq \text{MAX\_T}) \qquad (2)$$

Despite the extra effort, 55% of the teams have chosen to develop the stochastic features. Although they may have been motivated by the chance of getting a higher grade, this shows genuine interest on their side towards the project topics. The evaluation survey results presented in Sect. 5 confirm this intuition.

5   Evaluation Survey Results

We have invited students to fill an online evaluation survey[5] to assess their satisfaction level for the course and its effectiveness. Despite it being optional, 64% of all students who participated in the group assignment filled out the survey, whose results we comment in detail in the following. In some cases, results are compared with the ones previously presented by Askarpour and Bersani [3] to assess the evolution of the course with respect to its previous editions. About 75% of all students attended the course during their 1st M.Sc. year (fourth year of academic education according to the Italian system). For 6 students out of 10, this was an **optional** course, which is a reassuring indication given the historical low attendance that affects this course.

Concerning the students' attitude and expectations towards learning formal methods before attending the course, only 20% of them state that they had prior experience with formal methods. Moreover, Fig. 1a shows the students' self-assessed level of **confidence** for these topics *before* attending this course which amounts to an average of 2.33/5. Although this may be a physiological consequence of a student's lack of knowledge in a specific area before receiving education, we can consider the increase of confidence shown in Fig. 1b as a valuable achievement. The reported confidence level after attending the course is, indeed, 3.8/5 and, most importantly, no student chose a value lower than 3, which hints at a homogeneous improvement for the whole classroom. The reported reasons of low confidence unsurprisingly mention mathematical notation and lack of prior expertise as the main sources.

The questionnaire features questions specifically targeting the group assignment effectiveness. The results shown in Fig. 2 provide evidence that this is a

---

[5] Interested readers find the full set of questions at: https://bit.ly/3gAiHbS.

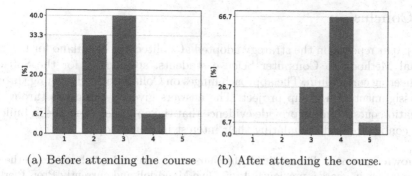

(a) Before attending the course   (b) After attending the course.

Fig. 1. Students self-assessed level of confidence ($[1-5]$) on course topics.

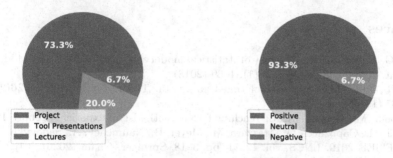

(a) Favourite part of the course.   (b) Impact on interest on course topics.

Fig. 2. Students replies targeting group assignment effectiveness.

successful strategy. Almost 75% of the students chose the project as the most appreciated part during the course. Moreover, the vast majority (93.3%) stated that working on the project **increased** their **interest** in course topics and, according to these results, no one lost interest because of the project. To complement the data in Fig. 2, 93.3% of the respondents also stated that the course stimulated their curiosity towards FM, and 4 people out of 10 said that they would consider a FM-related project for their Master Thesis (compared to the 1/10 ratio from two years ago [3]). Finally, the question about whether they would recommend the course to other students received an average score of 4.0/5.0, which is in line with the average of Politecnico courses (3.2/4.0).

Despite the favorable scenario, these results are not exempt from validity threats. Although most students enrolled in the course filled the survey, the original classroom size was meager, leading to only 25 replies. Furthermore, the survey carried out by Askarpour and Bersani in 2019 did not include specific questions about the project [3]. Therefore, we can only assess its effectiveness based on a single-year investigation. We will undoubtedly iterate the analysis for upcoming editions of the course to monitor future progress.

6   Conclusion

This paper reports on the strategy adopted at Politecnico di Milano for teaching Formal Methods to Computer Science students, specifically for the Software Engineering curriculum. The approach hinges on Collaborative Learning through the assignment of a group project. The answers given by students through an evaluation questionnaire provide evidence that the approach succeeds in building their confidence and stimulating their interest in course topics.

Acknowledgments. The credit for the course structure and syllabus goes to the official professors in charge, previously Prof. Dino Mandrioli and currently Prof. Pierluigi San Pietro.

References

1. Agha, G., Palmskog, K.: A survey of statistical model checking. ACM Trans. Model. Comput. Simul. (TOMACS) **28**(1), 1–39 (2018)
2. Alur, R., Dill, D.L.: A theory of timed automata. Theor. Comput. Sci. **126**(2), 183–235 (1994)
3. Askarpour, M., Bersani, M.M.: Teaching formal methods: an experience report. In: Bruel, J.-M., Capozucca, A., Mazzara, M., Meyer, B., Naumchev, A., Sadovykh, A. (eds.) FISEE 2019. LNCS, vol. 12271, pp. 3–18. Springer, Cham (2020). https://doi.org/10.1007/978-3-030-57663-9_1
4. Bareiss, R., Griss, M.: A story-centered, learn-by-doing approach to software engineering education. In: Proceedings of the 39th SIGCSE Technical Symposium on Computer Science Education, pp. 221–225 (2008)
5. Desel, J.: Teaching system modeling, simulation and validation. In: Winter Simulation Conference Proceedings, vol. 2, pp. 1669–1675. IEEE (2000)
6. Ghezzi, C., Mandrioli, D.: The challenges of software engineering education. In: Inverardi, P., Jazayeri, M. (eds.) ICSE 2005. LNCS, vol. 4309, pp. 115–127. Springer, Heidelberg (2006). https://doi.org/10.1007/11949374_8
7. Gibson, P., Méry, D.: Teaching formal methods: lessons to learn. In: 2nd Irish Workshop on Formal Methods 2, pp. 1–13 (1998)
8. Gleirscher, M., Marmsoler, D.: Formal methods in dependable systems engineering: a survey of professionals from Europe and North America. Empir. Softw. Eng. **25**(6), 4473–4546 (2020). https://doi.org/10.1007/s10664-020-09836-5
9. Hähnle, R., Bubel, R.: A hoare-style calculus with explicit state updates. Formal Methods in Computer Science Education, pp. 49–60 (2008)
10. Hinchey, M., Jackson, M., Cousot, P., Cook, B., Bowen, J.P., Margaria, T.: Software engineering and formal methods. Commun. ACM **51**(9), 54–59 (2008)
11. Korečko, Š, Sorád, J., Dudláková, Z., Sobota, B.: A toolset for support of teaching formal software development. In: Giannakopoulou, D., Salaün, G. (eds.) SEFM 2014. LNCS, vol. 8702, pp. 278–283. Springer, Cham (2014). https://doi.org/10.1007/978-3-319-10431-7_21
12. Laal, M., Laal, M.: Collaborative learning: what is it? Procedia-Soc. Behav. Sci. **31**, 491–495 (2012)
13. Larsen, K.G., Pettersson, P., Yi, W.: Uppaal in a nutshell, vol. 1, pp. 134–152. Springer-Verlag (1997)

14. Liu, S., Takahashi, K., Hayashi, T., Nakayama, T.: Teaching formal methods in the context of software engineering. ACM SIGCSE Bull. **41**(2), 17–23 (2009)
15. Mandrioli, D.: On the heroism of really pursuing formal methods. In: FME Workshop on Formal Methods in Software Engineering, pp. 1–5. IEEE (2015)
16. Parnas, D.L.: Really rethinking formal methods. Computer **43**(1), 28–34 (2010)
17. Parts, V., Teichmann, M., Rüütmann, T.: Would engineers need non-technical skills or non-technical competences or both? (2013)
18. Spichkova, M., Liu, H., Laali, M., Schmidt, H.W.: Human factors in software reliability engineering. arXiv preprint arXiv:1503.03584 (2015)
19. Spichkova, M., Zamansky, A.: Teaching of formal methods for software engineering. In: ENASE, pp. 370–376 (2016)
20. Tucker, A.B., Kelemen, C.F., Bruce, K.B.: Our curriculum has become mathphobic! In: Technical Symposium on Computer Science Education, pp. 243–247 (2001)

Lessons of Formal Program Design in Dafny

Ran Ettinger[✉]

Ben-Gurion University of the Negev, Beer Sheva, Israel
ranger@cs.bgu.ac.il

Abstract. Building on the long tradition of program derivation, whereby starting from a formal specification and progressing in small steps of refinement we end-up with correct executable code, this paper presents an approach for teaching that craft using the language and verifier Dafny. Some lessons from the first six years of teaching this material to final-year CS and SE undergraduate students are reported, with emphasis on the merits (and challenges) of using Dafny during live interactive sessions in the classroom.

Keywords: Refinement laws · Specification statement · Auto-active verification · Insertion sort

1 Introduction: About the Course

The textbook "Programming from Specifications" (*PfS*) by Carroll Morgan [8] introduces a student into the world of program derivation in a smooth and formal way. Equipping the novice formal programmer with motivation, some logical background on the predicate calculus, and elementary means known as *laws of refinement* for developing correct imperative programs (using a programming notation based on Dijkstra's guarded commands [2]), the book dedicates its tenth chapter to presenting a case study for developing a first program with nested loops: *insertion sort*.

The insertion sort case study is reformulated, in this paper, using the language and verifier Dafny [5]. This acts as a basis for reporting on some experiences from the first six years of teaching a substantial subset of the *PfS* material in a course entitled "correct-by-construction programming" (*ccpr*[1]). This is an elective course given to final-year CS and SE undergraduate students at Ben-Gurion University. Following *PfS*, the course teaches how to design algorithms and programs that are guaranteed to meet their specification. Starting with a mathematical description of the program's requirements, the course presents a formal method for turning such specifications into actual code, in a stepwise approach known as refinement. Techniques of algorithm refinement are presented, for the derivation of loops from invariants, as well as recursive procedures.

---

[1] Course website: https://www.cs.bgu.ac.il/~ccpr191.

© Springer Nature Switzerland AG 2021
J. F. Ferreira et al. (Eds.): FMTea 2021, LNCS 13122, pp. 84–100, 2021.
https://doi.org/10.1007/978-3-030-91550-6_7

The developed algorithms are typically very short, but challenging, as we aim to construct correct and efficient code. The programming throughout this course is done in the language Dafny, using its integration into Microsoft Visual Studio [7]. This environment enables the annotation of programs with their specifications. Moreover, it includes an automatic verifier, such that a program can be executed only after its functional correctness has been established (with some potential exceptions to be discussed later in the paper). A switch to Visual Studio Code is currently scheduled for the next iteration of the course, along with some other changes, including the adoption of SPARK/Ada (that has commenced in the 2020 iteration of the course) with its GNATprove verifier and GNAT Programming Studio (GPS) IDE.

The main textbook of this course is "Programming from Specifications" by Carroll Morgan [8] and related material can be found in further sources, including [1–4,6]. A subset of Morgan's laws of refinement is being introduced gradually in the first third of the course[2], through live sessions of program derivation in class. Most programs are taken from Morgan's book. For example, in the first few weeks, we learn how to develop a loop, correctly, by deriving iterative programs for computing a Fibonacci number, the factorial of a natural number, the non-negative floor of the square root of a given natural number (through linear and then binary search). This is followed by the development algorithms to search for an element in a read-only array. Equipped with a basic familiarity of how to design and implement iterative algorithms, using loop invariants, in small and provably-correct steps of refinement, we are ready for the first bigger case study. This will be our first program to update the contents of a given array, and our first derivation of a nested loop.

2 Lessons 10–12: Insertion Sort

The *ccpr* course has been given at BGU in each Fall semester since October 2013. It is made of two sessions a week, of two hours each, over a period of 13 weeks. Approaching the middle of the semester we typically dedicate three sessions to the *PfS* case study of insertion sort [8, Chapter 10]. In the 2019 iteration of the course, reported here, these three sessions commenced on the 10th lecture of the semester. The complete derivation comprising 11 steps of refinement, is given in detail in Figs. 1, 2, 3, 4, 5, 6, 7 and 8. The eventual code is shown in Fig. 9. (As I make frequent references to line numbers of program elements, in what follows, it may be helpful, if possible, to have two copies of the paper in front of you. This is how I evaluate homework submissions: reading the call to a method or lemma in one part of the program, and quickly browsing in the other copy to study its specification, comparing it to the expected specification according to the presently exercised law of refinement).

2.1 Specification for a Sorting Algorithm

A possible specification for sorting an array of integers in a non-decreasing order is shown on lines 1–10 of the program in Fig. 1. Referring to the predicate

---

[2] https://www.cs.bgu.ac.il/~ccpr191/Laws_Of_Refinement.

```
1   predicate Sorted(q: seq<int>)
2   {
3       ∀ i,j • 0 ≤ i ≤ j < |q| ⟹ q[i] ≤ q[j]
4   }
5
6   method InsertionSort(a: array<int>, ghost A: multiset<int>)
7       requires multiset(a[..]) = A
8       ensures Sorted(a[..])
9       ensures multiset(a[..]) = A
10      modifies a
11  {
12      // Step 1: introduce local variable + strengthen postcondition
13      var i := InsertionSort1(a, A);
14      StrongerPostcondition1(a,i,A);
15  }
16
17  predicate Inv1(a: array<int>, i: nat, A: multiset<int>) reads a
18  {
19      i ≤ a.Length ∧
20      Sorted(a[..i]) ∧
21      multiset(a[..]) = A
22  }
23
24  lemma StrongerPostcondition1(a: array<int>, i: nat, A: multiset<int>)
25      requires Inv1(a,i,A) ∧ i = a.Length
26      ensures Sorted(a[..]) ∧ multiset(a[..]) = A
27  {}
28
29  method InsertionSort1(a: array<int>, ghost A: multiset<int>)
30          returns (i: nat)
31      requires A = multiset(a[..])
32      ensures Inv1(a,i,A) ∧ i = a.Length
33      modifies a
```

Fig. 1. A specification for sorting along with a first step of refinement, reflecting a design for the anticipated outer loop.

Sorted from lines 1–4, the postcondition on line 8 expresses the expectation that when exiting InsertionSort, the sequence of elements stored in the given array (denoted a[..] in Dafny) will be in a non-decreasing order. The fact that this sequence is a permutation of the original contents of the array is expressed here as a combination of the precondition and postcondition on lines 7 and 9 respectively, using the additional parameter A. Being a ghost parameter, A acts here as a logical constant, storing the bag of values in the given array; in Dafny we achieve this through the type multiset (line 6) and the operator with the same name (lines 7 and 9) that collects the bag of values from the sequence of numbers stored in the array. Following Morgan's presentation, we typically start the session by considering how to specify the requirements of sorting, highlighting the need to express the fact that the eventual array contents must be a permutation of the original: in the absence of the postcondition on line 9, a "correct" implementation could possibly set all elements of the array with the value 7. As this session provides a first example of an algorithm that updates the contents of a heap object, we see here for the first time the modifies clause, on line 10. In class, I typically start without it, showing how Dafny complains correctly about an assignment to the array, saying it "may update an array element not in the enclosing context's modifies clause". In Morgan's terminology of a specification statement, the key ingredients here (aside from the definitions of the variables and the predicate) are the frame (line 10), the precondition (line 7), and the postcondition (lines 8–9). Morgan's original specification is slightly cleaner in that it expresses the multiset property of lines 7 and 9 as an invariant of the program (with respect to the contents of the array a and the constant A); to the best of my knowledge, this is not currently supported by Dafny.

Some students feel inclined to add a precondition stating that the array is not empty or that it has at least two elements. They are correct in their observation that below two elements there is no need to do anything. But I try to make it clear that it is against the rules of our game to change the specification. This specification should be seen as a binding contract, between the programmer (them and me, in the classroom) and our invisible client. Should it indeed be helpful to assume the array has at least two elements, they could always start the implementation with an alternation, asking if it has at least two elements in an if-statement; the then part will call a method whose precondition can explicitly state that the array has at least two elements. I also teach them never to leave an if-statement with no else part. Instead, we show that the else is redundant using the *skip command* refinement law. Morgan's approach to alternation is more general, explicitly requiring that the precondition will entail the disjunction of all guards of a guarded command.

In contrast to the common practice of (a-posteriori) verification of existing code, we shall develop the code through a process of stepwise refinement. In class, as an exercise, we sometimes agree in advance to aim at the development of specific code. Still it is important to keep in mind the spirit of correct-by-construction programming, with the code and proof being developed side by side. At the end of the process, we will have two versions of the code: the inlined version as shown later in this paper (Fig. 9), and the complete version, comprising 11 methods (Figs. 1, 2, 3, 4, 5, 6, 7 and 8). One advantage of the complete version, in spite of its length, lies in its persistent documentation of the refinement process. A student who missed that class or was unable to follow the interactive development, would ideally be able to reconstruct the full process from the published final version[3].

2.2   Refinement Steps 1–5: The Outer Loop

In the first five steps of refinement we develop a loop for successive insertion of elements into their sorted location in the prefix of the array. This process acts as a derivation of a precise specification for the anticipated Insert operation, to move the next element into its correct (sorted) location in the prefix to its left. As in *PfS*, this example is the first in the course in which we end-up with a nested loop. The code for the nested loop itself will be developed subsequently, in refinement steps 6–11 below, starting from the derived specification for the Insert operation.

---

[3] Final version of the insertion sort algorithm from the 2019 iteration of the *ccpr* course (including detailed proofs for the human reader): https://www.cs.bgu.ac.il/~ccpr191/wiki.files/CCPR191-InsertionSort-complete-10Dec18.dfy.

The first step of refinement, shown in Fig. 1, takes the original specification of `InsertionSort` (lines 6–10) and implements it by providing a method body (lines 11–15). This is our course's form of expressing refinement in Dafny. Whenever the refined program involves yet-to-be-implemented specification statements, additional methods are being specified (here `InsertionSort1` on lines 29–33), and can already be invoked (line 13), leaving their implementation for later refinement steps. This first step introduces the local variable i, to act as a loop index, and strengthens the postcondition in anticipation for the loop. For this step to be correct, we have an obligation to prove that the new postcondition (line 32) is indeed stronger than the original postcondition (lines 8–9). A convenient way to document such proof obligations in Dafny is through the specification of a lemma (lines 24–26). The generation of this specification is taught as a mechanical process of copying-and-pasting: the new postcondition (line 32) acts as the lemma's precondition (line 25); the older postcondition (lines 8–9) acts as the lemma's postcondition (line 26); and all relevant variables are sent as parameters.

In technical terms, the lemma acts as a ghost method, with no side effect, and in this case with value parameters only. Seeing an invocation of the lemma (line 14), Dafny takes the responsibility to verify that the lemma's precondition holds; in this case Dafny trusts that it does hold, as the call immediately follows the invocation of `InsertionSort1` (line 13) whose postcondition is, by design, the lemma's precondition. And then Dafny assumes that on return from the lemma, its postcondition holds, which is again by design the original postcondition of `InsertionSort`. And hence Dafny has no reason to complain that the postcondition of `InsertionSort` might not hold. In this sense, Dafny trusts its user to prove at some point in the development that the lemma is correct. In class we sometimes leave the lemma unproved at first, just as we do with specifications of further methods, leaving their development for a later step. In this case, however, Dafny gets convinced of the correctness of this lemma with no need for proof. This is the meaning of the lemma's empty body (line 27). Had Dafny been unable to prove correctness of the lemma, it would have complained that a postcondition of the lemma might not hold.

What is it that makes the lemma correct in this case? Following *PfS*, the designed loop invariant `Inv1` expresses the expectation that the index i does not exceed the size of the array (line 19), and that the first i elements are sorted (line 20). The fact the loop invariant and the negation of its guard hold (line 25), ensures that the first `a.Length` elements (hence the entire array) are sorted. And the second conjunct of the lemma's postcondition directly follows from the third conjunct of the loop invariant (line 21), stating that the multiset of values in the array is indeed the expected multiset, as stored in `A`.

In logical terms, such a lemma, formulated with input parameters only, expresses what Morgan refers to as entailment [8]: the expectation that for all values of the input parameters, according to their types, the result of the lemma's precondition implies the result of the postcondition. In other words, for all values on which the precondition holds, the postcondition must hold too. (Output parameters from a lemma add an existential portion to the formula, that there exist values of these parameters, for which the implication holds).

At the end of this first step of refinement, as said, we are left to continue the development by implementing method `InsertionSort1`. Its specification has been derived by that of `InsertionSort` with two differences: the postcondition has been strengthened, as discussed above, and the frame has been extended (line 30) to accommodate modifications to the value of the loop index, i. Using output parameters from methods through the `returns` construct (line 30), along with the `modifies` clause (line 33) is our way of expressing Morgan's frame in Dafny. And adding i to the frame here is a direct effect of Morgan's refinement law called *introduce local variable*.

```
29    method InsertionSort1(a: array<int>, ghost A: multiset<int>)
30          returns (i: nat)
31       requires A = multiset(a[..])
32       ensures Inv1(a,i,A) ∧ i = a.Length
33       modifies a
34    {
35       // Step 2: sequential composition + contract frame
36       i := InsertionSort2a(a,A);
37       i := InsertionSort2b(a,i,A);
38    }

40    method InsertionSort2a(a: array<int>, ghost A: multiset<int>)
41          returns (i: nat)
42       requires A = multiset(a[..])
43       ensures Inv1(a,i,A)

55    method InsertionSort2b(a: array<int>, i0: nat, ghost A: multiset<int>)
56          returns (i: nat)
57       requires Inv1(a,i0,A)
58       ensures Inv1(a,i,A) ∧ i = a.Length
59       modifies a
```

Fig. 2. Sequential composition: establish the invariant first and only then get to the loop.

In a second step of refinement, as further preparation for the loop, we decompose the implementation (of method `InsertionSort1`) into a sequence of two operations, as can be seen on lines 36–37 of Fig. 2. The first operation will establish the loop invariant (as can be witnessed in its specification on line 43), and the second operation will be the loop itself. The postcondition of the first operation in a sequential composition, according to the simplest version of this law of refinement, may act as the precondition to the second operation. Whenever we aim for a loop, as we do here, we choose the loop invariant to be this property (lines 43 and 57). Note however that in the precondition of `InsertionSort2b` we refer to i0 rather than i. This is our way of implementing parameter passing to variables in the frame: according to the common convention, we append the digit 0 to the name of a variable whose initial value is required and whose value may be modified in the method. In contrast to Morgan, each refinement may introduce new scopes for variables, and accordingly, the initial variable (such as i0 here) is a genuine parameter, not merely a (ghost) logical constant. While these variables and assignment statements could be seen to have negative impact on the performance of the derived implementation, it is good to recall that by collecting the code at the end of the refinement process, inlining all method bodies, such variables can be removed.

As we anticipate that modifications to the array's contents will be performed only in the loop body, we express this decision explicitly by removing a from the frame of the initialization method, leaving only the loop index in its frame (line 41). This is a refinement step known as *contract frame*.

```
40    method InsertionSort2a (a: array<int>, ghost A: multiset<int>)
41          returns (i: nat)
42       requires A = multiset (a [..])
43       ensures Inv1(a,i,A)
44    {
45       // Step 3: assignment
46       LemmaInsertionSort2a(a,A);
47       i := 0;
48    }
49
50    lemma LemmaInsertionSort2a(a: array<int>, A: multiset<int>)
51       requires A = multiset (a [..])
52       ensures Inv1(a,0,A)
53    {}
```

Fig. 3. A first example of assignment: the proof obligation resembles the original specification, with substitution (of the assignment's LHS by its RHS) performed on the postcondition.

In a third step of refinement, shown in Fig. 3, we choose to implement InsertionSort2a, presenting a first assignment statement, to initialize the outer loop index. The proof obligation of an assignment statement is expressed as a lemma specification (lines 50–52). The lemma states that the precondition entails a modified version of the postcondition, obtained by substituting the assignment's left-hand side with the corresponding right-hand side. Here, starting with a copy of the postcondition of InsertionSort2a, the loop index i has been substituted by 0 in the lemma's postcondition (line 52, compared to line 43). The lemma's precondition (line 51) in such cases remains unchanged (as in line 42). As the correctness of this lemma is proved by Dafny with no difficulties, we *implement* it immediately with an empty body (line 53). This is the first step that introduces no further specifications: the refined version is executable code.

Shown in Fig. 4, the fourth step of our refinement session introduces the outer loop. As the first refinement of a specification with initial variables, we see here for the first time a convention of copying the initial value to the output variable (line 61). As in some of the previously demonstrated laws, *iteration* requires no proof obligation. Instead, we must be sure to start with a specification that expresses the loop invariant in its precondition (line 57, using the initial variable i0) and its postcondition must be phrased as a conjunction of the loop invariant and the negation of the loop guard (line 58). Following Morgan's iteration law, the specification of the loop body should express the loop invariant and the loop guard in its precondition (line 73, again with initial variables), and the postcondition (line 74) must involve both the loop invariant and an indication that the loop variant is strictly decreasing, yet not below some lower bound (typically chosen to be 0); the frame of the loop body remains unchanged (lines 72 and 75).

In class, it is helpful to see how commenting out the first conjunct of the loop body's postcondition on line 74 leads to an error reported on line 64: "This loop invariant might not be maintained by the loop". Alternatively, commenting out the *second* conjunct on line 74 (involving termination of the loop) leads to an error reported on line 63, stating that the "decreases expression might not decrease". In contrast to that, I sometimes forget to include the guard in the loop body's precondition (line 73), and we get no error; only later in the development we come to notice its absence and learn to appreciate its significance: such a specification would be *infeasible* as the result of the precondition not being strong enough here is that *there exists no value* for the output parameter i that satisfies the postcondition.

```
55    method InsertionSort2b(a: array<int>, i0: nat, ghost A: multiset<int>)
56          returns (i: nat)
57       requires Inv1(a,i0,A)
58       ensures Inv1(a,i,A) ∧ i = a.Length
59       modifies a
60    {
61       i := i0;
62       // Step 4: iteration
63       while i ≠ a.Length
64          invariant Inv1(a,i,A)
65          decreases a.Length-i
66       {
67          i := InsertionSort3(a, i, A);
68       }
69    }
70
71    method InsertionSort3(a: array<int>, i0: nat, ghost A: multiset<int>)
72          returns (i: nat)
73       requires Inv1(a,i0,A) ∧ i0 ≠ a.Length
74       ensures Inv1(a,i,A) ∧ 0 ≤ a.Length-i < a.Length-i0
75       modifies a
```

Fig. 4. The outer loop: the specification of the loop body is mechanically derived with copies of the invariant (twice), the guard, and the variant function.

The loop body is expected to make two changes: it should increment the loop index and it must insert the next element into its sorted location in the growing prefix of the array. Focusing on the loop index first, our fifth step of refinement, shown in Fig. 5, reflects a decision to increment i at the end of the loop body. This step, known as *following assignment*, is quite simple to perform. The specification of method Insert (lines 83–86) reflects the expectations from the remaining part of the loop body (line 79, to be followed both in the program text and execution time by the assignment to i on line 80) is nearly identical to the specification of the loop body (lines 71–75), with only a few differences.

The single update due to the *following assignment* law causes each reference of i in the postcondition to be substituted by i+1 (line 85). Since we anticipate no further changes to i, we remove it from the frame, causing one subsequent change, replacing i0 by i. It is important to note the order here: first substitution (of i only, not of i0) then rename of i0 back to i. At the end of this modification, the variant-related part of the postcondition becomes trivially true: the a.Length-i < a.Length-i0 is now the obviously correct condition a.Length-(i+1) < a.Length-i and the 0 <= a.Length-i is now 0 <= a.Length-(i+1), which is equivalent to the first conjunct of the loop invariant (line 19 on Fig. 1), applied here in line 85 to i+1. So we do not repeat this (by-now-redundant) part in the postcondition of Insert (line 85). Indeed, it is frequently the case that this combination of *following assignment* and *contract frame* makes the variant portion of the loop body's postcondition trivially true.

```
71    method InsertionSort3(a: array<int>, i0: nat, ghost A: multiset<int>)
72         returns (i: nat)
73        requires Inv1(a,i0,A) ∧ i0 ≠ a.Length
74        ensures Inv1(a,i,A) ∧ 0 ≤ a.Length-i < a.Length-i0
75        modifies a
76    {
77        i := i0;
78        // Step 5: following  assignment + contract frame
79        Insert(a,i,A);
80        i := i+1;
81    }
82
83    method Insert(a: array<int>, i: nat, ghost A: multiset<int>)
84        requires Inv1(a,i,A) ∧ i ≠ a.Length
85        ensures Inv1(a,i+1,A)
86        modifies a
```

Fig. 5. Updating the loop index and deriving a specification for the remaining computation (the insert operation).

2.3   Refinement Steps 6–10: The Inner Loop

In the second session dedicated to insertion sort, we get to the development of the inner loop. This is more challenging, compared to the derivation of the outer loop, mostly due to the need to change the contents of the array. Accordingly, the loop invariant, the proof obligations, and the proof itself might all be more complicated. The first step in the development of this inner loop is shown in

```
83    method Insert(a: array<int>, i: nat, ghost A: multiset<int>)
84        requires Inv1(a,i,A) ∧ i ≠ a.Length
85        ensures Inv1(a,i+1,A)
86        modifies a
87    {
88        // Step 6: introduce local variable + strengthen postcondition
89        var j := Insert1(a,i,A);
90        StrongerPostcondition2(a,i,j,A);
91    }
92
93    predicate SortedExceptAt(q: seq<int>, k: nat)
94    {
95        ∀ i,j • 0 ≤ i ≤ j < |q| ∧ i ≠ k ∧ j ≠ k ⟹ q[i] ≤ q[j]
96    }
97
98    predicate Inv2(q: seq<int>, i: nat, j: nat, A: multiset<int>)
99    {
100       j ≤ i < |q| ∧
101       SortedExceptAt(q[..i+1],j) ∧
102       (∀ k • j < k ≤ i ⟹ q[j] < q[k]) ∧
103       multiset(q) = A
104   }
105
106   predicate method InsertionGuard(a: array<int>, i: nat, j: nat,
107                 ghost A: multiset<int>)
108       requires Inv2(a[..],i,j,A)
109       reads a
110   {
111       1 ≤ j ∧ a[j-1] > a[j]
112   }
113
114   lemma StrongerPostcondition2(a: array<int>, i: nat, j: nat, A: multiset<int>)
115       requires Inv2(a[..],i,j,A) ∧ ¬InsertionGuard(a,i,j,A)
116       ensures Inv1(a,i+1,A)
117   {}
118
119   method Insert1(a: array<int>, i: nat, ghost A: multiset<int>)
120           returns (j: nat)
121       requires Inv1(a,i,A) ∧ i ≠ a.Length
122       ensures Inv2(a[..],i,j,A) ∧ ¬InsertionGuard(a,i,j,A)
123       modifies a
```

Fig. 6. Preparation for the insertion loop, defining a loop invariant and a guard, this time in its own *predicate method*, aiming for enhanced clarity of annotations.

Fig. 6. Recalling the definition of the outer loop invariant (Inv1 on lines 17–22 of Fig. 1), the specification of Insert (lines 83–86) could be interpreted as saying that given a state in which the first i elements in an array are sorted and there is at least one more element to sort, namely a[i], we wish to *insert* it into its correct location such that the first i+1 elements will be sorted. (The specification also says that we must maintain the existing elements in the array; confining array modifications to swapping pairs of elements will satisfy this requirement.)

To explore the definition of the inner loop invariant (Inv2 on lines 98–104, using an additional predicate on lines 93–96) and the definition of the loop guard, expressed in its own predicate method (lines 106–112) such that it can be used both in executable code and in annotations, it is helpful to consider the specification of lemma StrongerPostcondition2 (lines 114–116, invoked on line 90). Fortunately again, this lemma is proved by Dafny, hence the empty curly braces (line 117) for its proof. In words, following Morgan's design, this is indeed true since when the loop invariant holds and the loop guard does not (line 115), the state is such that the first i+1 elements are sorted except at index j (line 101) and a[j] is sorted (among the first i+1 elements) too; the latter is true thanks to a healthy combination of the loop invariant and the negation of the guard: the inserted element, at location j, is guaranteed to be sorted to its right thanks to the loop invariant (line 102) and it is guaranteed to be sorted to its left thanks to the negation of the guard (from line 111), since at that state either it is the leftmost element, or it is not smaller than the element to its left, which along with the loop invariant (line 101 again, taken together with line 100) means that the inserted element (at index j) is indeed greater-or-equal all elements to its left.

As can be guessed by reading the loop guard, we are aiming for a loop body that repeatedly swaps the inserted element with the element to its left, until it reaches its expected location (either when there are no more elements on its left, in case it is the smallest, or when the element to its left is not larger). It is instructive to see here how the loop invariant records key properties from the history of the computation. Failing to record in the loop invariant (line 102) the fact that at each iteration, and most importantly at the end of the last (line 122), all the *previously* considered elements which are *currently* placed to the right of the inserted element are greater than the inserted element. Commenting out this property, removing it from the loop invariant (line 102), immediately leads to failure in the proof attempt of lemma `StrongerPostcondition2` (line 117). In homework assignment submissions, it is not uncommon to find a comment attached to such an unproved lemma, *waving hands* about what is expected to be true at the point of lemma invocation (line 90 in this case, after the loop). My response in such cases is that the separation of concerns in our proof method is such, that the lemma reflects a logical property that stands by itself; if proven correct (along with separate proofs for all the other obligations), it guarantees that the program satisfies its specification; yet when the lemma by itself is logically incorrect, I simply try to provide a counterexample, in this case with a smaller element to the right of the inserted one; students might argue that my counterexample does not make sense, and that at the end of the loop we will never find such smaller elements to the right of the inserted element; and indeed the fact that we are unable to prove correctness does not necessarily imply that our code is incorrect; it simply means we need to try harder, for example by strengthening the loop invariant, recording there more information from the history of the computation. To such students, it may be helpful to see here on Fig. 6 that the question of whether the loop invariant and the negation of its guard imply for all states that the postcondition of the loop holds can be addressed even before we have implemented the loop.

The development of the inner loop itself is documented in the steps 7–10 of our refinement scenario, as shown in Fig. 7, culminating in a specification for the final operation, of swapping two adjacent elements of the array (lines 173–176). It follows the same line as steps 2–5 of the outer loop: *sequential composition* with *contract frame, assignment, iteration*, and then *following assignment* with *contract frame*. With Morgan's rich repertoire of refinement laws there is a variety of paths for deriving the same eventual code. Indeed, in class we cover some

```
119   method Insert1(a: array<int>, i: nat, ghost A: multiset<int>)
120         returns (j: nat)
121      requires Inv1(a,i,A) ∧ i ≠ a.Length
122      ensures Inv2(a[..],i,j,A) ∧ ¬InsertionGuard(a,i,j,A)
123      modifies a
124   {
125      // Step 7: sequential composition + contract frame
126      j := Insert2a(a,i,A);
127      j := Insert2b(a,i,j,A);
128   }
129
130   method Insert2a(a: array<int>, i: nat, ghost A: multiset<int>)
131         returns (j: nat)
132      requires Inv1(a,i,A) ∧ i ≠ a.Length
133      ensures Inv2(a[..],i,j,A)
134   {
135      // Step 8: assignment
136      LemmaInsert2a(a,i,A);
137      j := i;
138   }
139
140   lemma LemmaInsert2a(a: array<int>, i: nat, A: multiset<int>)
141      requires Inv1(a,i,A) ∧ i ≠ a.Length
142      ensures Inv2(a[..],i,i,A)
143   {}
144
145   method Insert2b(a: array<int>, i: nat, j0: nat, ghost A: multiset<int>)
146         returns (j: nat)
147      requires Inv2(a[..],i,j0,A)
148      ensures Inv2(a[..],i,j,A) ∧ ¬InsertionGuard(a,i,j,A)
149      modifies a
150   {
151      j := j0;
152      // Step 9: iteration
153      while InsertionGuard(a,i,j,A)
154         invariant Inv2(a[..],i,j,A)
155         decreases j
156      {
157         j := Insert3(a,i,j,A);
158      }
159   }
160
161   method Insert3(a: array<int>, i: nat, j0: nat, ghost A: multiset<int>)
162         returns (j: nat)
163      requires Inv2(a[..],i,j0,A) ∧ InsertionGuard(a,i,j0,A)
164      ensures Inv2(a[..],i,j,A) ∧ j < j0
165      modifies a
166   {
167      j := j0;
168      // Step 10: following assignment + contract frame
169      Swap(a,i,j,A);
170      j := j - 1;
171   }
172
173   method Swap(a: array<int>, i: nat, j: nat, ghost A: multiset<int>)
174      requires Inv2(a[..],i,j,A) ∧ InsertionGuard(a,i,j,A)
175      ensures Inv2(a[..],i,j-1,A)
176      modifies a
```

Fig. 7. Four steps of refinement in the development of the inner loop, deriving a specification for the swap operation. Note the similarity to steps 2–5.

more laws, but still quite a small subset of the original catalog. (The additional laws we *do* cover include *alternation, skip command, leading assignment* and *weaken precondition.*)

```
173    method Swap(a: array<int>, i: nat, j: nat, ghost A: multiset<int>)
174        requires Inv2(a[..],i,j,A) ∧ InsertionGuard(a,i,j,A)
175        ensures Inv2(a[..],i,j-1,A)
176        modifies a
177    {
178        // Step 11: assignment
179        LemmaSwap(a,i,j,A);
180        a[j-1],a[j] := a[j],a[j-1];
181    }
182
183    lemma LemmaSwap(a: array<int>, i: nat, j: nat, A: multiset<int>)
184        requires Inv2(a[..],i,j,A) ∧ InsertionGuard(a,i,j,A)
185        ensures Inv2(a[..][j-1 := a[j]][j := a[j-1]],i,j-1,A)
186    {}
```

Fig. 8. One last step of refinement, swapping the inserted element with the (larger) array item on its left. Note the "sequence assignment" in the proof obligation.

2.4   A Final Step of Refinement: Swapping Adjacent Array Elements

Our last step of refinement, performed in the third and final session dedicated to insertion sort, is shown in Fig. 8. This final step is particularly interesting in the way its proof obligation uses sequence assignment in the lemma specification. It is for the purpose of this substitution that we expressed the inner loop invariant as a predicate that expects a sequence rather than an array of integers, as one of its parameters (line 98 on Fig. 6). According to the proof obligation for the *assignment* law of refinement, note how the specification of LemmaSwap is similar to that of Swap, except that the frame is empty, and in the postcondition (line 185), only the a[..] has been substituted at two locations, based on the LHS of the multiple assignment, with values from its RHS. Magically, as was the case with all prior lemma specifications in this derivation of insertion sort, this lemma too is proved by Dafny (line 186). Indeed, when the first i+1 elements are sorted except at j, and a[j] is greater-or-equal all elements to its right, yet it is smaller than a[j-1], swapping them (a[j] and a[j-1]) generates a sequence in which the first i+1 elements are sorted except at j-1 and a[j] in its new location is indeed smaller-or-equal all elements to its (new) right, as these are the elements right of j as well as j-1, now at location j.

In class, it is actually only in this third session that we transform the inner loop invariant to take a sequence, rather than the array, as a parameter. This enables the expression of the proof obligation for the swap assignment using sequence assignments. In retrospect, sending a sequence rather than the array could be a more appropriate choice for the outer loop invariant too. This way, there would be no need to explain Dafny's **reads** frame (line 17 in Fig. 1), which is not present in Morgan's approach.

One more transformation we typically perform on the third session is of the guard of the inner loop. Following Morgan, we initially express this guard using an existential quantifier (that at least one of the first j elements of the array is larger than the inserted value), and at this stage we replace it with the more efficient guard as shown here in the paper. We use a lemma to demonstrate that when the loop invariant holds, these two formulations of the guard are equivalent.

In conclusion of this session, we observe that 11 steps of refinement were performed, developing executable code that is now scattered in 11 methods and one predicate method. Inlining these methods, in order to collect the code, would yield the version shown in Fig. 9.

```
method {: verify false} InsertionSort_TheCode(a: array<int>,
        ghost A: multiset<int>)
    requires multiset(a[..]) = A
    ensures Sorted(a[..])
    ensures multiset(a[..]) = A
    modifies a
{

    var i := 0;
    while i ≠ a.Length
    {
        var j := i;
        while 1 ≤ j ∧ a[j-1] > a[j]
        {
            a[j-1],a[j] := a[j],a[j-1];
            j := j-1;
        }
        i := i+1;
    }
}
```

Fig. 9. Collecting the correct-by-construction code at the end of the refinement process.

In the version of insertion sort we have just completed developing, as it turns out, we were somewhat lucky that each lemma was proved with no need for manual intervention. To appreciate this, suppose we were to define the predicate SortedExceptAt not as we did (on lines 93–96, Fig. 6), but rather in the following equivalent way: k < |q| && Sorted(q[..k]+q[k+1..]). As a result, we would get errors both in lemma StrongerPostcondition2 and in LemmaSwap, stating that "A postcondition might not hold". In such cases I encourage my students to spend some time in trying to prove correctness, yet not too much time. The official *order* is *not to fight* Dafny, as we do not learn how Dafny works. We return to discuss this challenge in the next section, reporting on the final homework assignment for this course.

3   Assessment

The final grade in the 2019 iteration of the *ccpr* course was determined by one homework assignment (20%), a must-pass midterm examination (20%), and a final assignment (60%). The assignments were performed by teams of at most three members. The first assignment[4] involved two exercises: (1) binary search, and (2) search for two elements in a sorted sequence of integers whose sum is a given number. The midterm examination was (for the third year running) a multiple-choice quiz[5]. The final assignment[6] involved three exercises: (1) merge sort; (2) inserting an element to a maximum-heap data structure; and (3) inserting an element into a binary-search tree. Of those, the `HeapInsert` seemed to be the trickiest. Here is the specification for this exercise:

```
predicate AncestorIndex(i: nat, j: nat) decreases j−i
{
    i = j ∨ (j > 2*i ∧
    ((AncestorIndex(2*i+1, j) ∨ AncestorIndex(2*i+2, j))))
}


predicate hp(q: seq<int>, length: nat)
    requires length ≤ |q|
{
    ∀ i,j • 0 ≤ i < length ∧
    0 ≤ j < length ∧ AncestorIndex(i, j) ⟹ q[i] ≥ q[j]
}


method HeapInsert(a: array<int>, heapsize: nat, x: int)
    requires heapsize < a.Length
    requires hp(a[..], heapsize)
    ensures hp(a[..], heapsize+1)
    ensures multiset(a[..heapsize+1]) = multiset(old(a[..heapsize])+[x])
    modifies a
```

The `HeapInsert` challenge was not given in isolation. The most involved program we develop during the semester is based on one more case study from *PfS*, for Heap Sort [8, Chapter 12]. Our complete solution to Heap Sort[7] contains nearly 500 lines of non-blank-non-comment executable code and annotations. Most of the ingredients for deriving a correct heap-insert algorithm were available in the heapsort solution; the goal of this exercise was to encourage the students to read the complete solution more closely; yet the development of fully verified solutions was beyond my expectations. Accordingly, the assignment description included the following text: "The submitted programs are expected to compile and verify with no errors (except perhaps for lemma specifications annotated with a {:verify false}, whose body is left empty) in Dafny 2.2.0. The correctness of all your non-proved ({:verify false}) lemma specifications should be made clear by a verbal comment, explaining why for all values of its parameters (according to their types), if the lemma's precondition holds then its postcondition must clearly hold too. Recall that for the lemma to be correct, this form of logical implication MUST hold by itself, independently of properties known to the reader from any other part of the program. Please note that

---

[4] https://www.cs.bgu.ac.il/~ccpr191/Assignments/Assignment_1.

[5] https://www.cs.bgu.ac.il/~ccpr191/Previous_Exams.

[6] https://www.cs.bgu.ac.il/~ccpr191/Assignments/Final_Assignment.

[7] https://www.cs.bgu.ac.il/~ccpr191/wiki.files/CCPR191-HeapSort-complete-30Dec18.dfy.

some of the properties required for completion of the development might be very difficult to prove in a formal way (as can be witnessed for example in the published `HeapSort` solution). In each such case you are indeed highly encouraged to formulate an appropriate lemma, explain to the human reader the reason for its correctness, and then leave the lemma unverified in the form stated above."

Table 1. Levels of success: an algorithm to insert an element into a maximum-heap.

Success level of correctness proof	SGs
Fully verified	8
Fully verified inconsistently: on occasion, the proof of one lemma fails	1
Perfectly convincing {:verify false} lemma specifications	0
Mostly well-argued {:verify false} lemma specifications	6
Badly-argued (probably correct) {:verify false} lemma specifications	3
Logically-incorrect {:verify false} lemma specifications	9
Seemingly correct code with {:verify false} methods	2
Incorrect implementation	3
Did not submit a solution to this portion of the final assignment	2

Table 1 provides a summary of the level of correctness and proof achieved by the 78 course participants, who teamed-up as 34 Submission Groups (SGs). I was encouraged and impressed by the fact that 8 submissions were fully verified. The 9 submissions with logically-incorrect lemma specifications show that there is certainly room for improvement, in my teaching. And perhaps more importantly, I would hope to improve the approach—possibly adopting ideas from Leino's "Program Proofs" (draft) book [6]—in a way that will help upgrade the 6 submissions on the fourth line to the empty third line. I believe that the simpler it would become, for the students, to provide proofs in which the only unproved properties will be easy to explain, to the human reader, some of the students in the 8 teams of the top row would settle for that third row. This will have saved them the time and energy of "fighting" with a theorem prover.

Performance of students on the heap-insert exercise, as reported above, may hopefully raise some optimism: perhaps it is not too late to introduce students to formal methods on the final year of their undergraduate studies (even though it is definitely advisable to start much earlier [9]), and hopefully new generations of practitioners (and of teachers) with skill and experience in formal program design could be raised this way.

References

1. Backhouse, R.: Program Construction: Calculating Implementations from Specifications. Wiley, New York (2003)
2. Dijkstra, E.W.: A Discipline of Programming. Prentice-Hall, Hoboken (1976)
3. Gries, D.: The Science of Programming. Springer, Heidelberg (1987)
4. Kaldewaij, A.: Programming: The Derivation of Algorithms. Prentice-Hall Inc., Upper Saddle River (1990)
5. Leino, K.R.M.: Dafny: an automatic program verifier for functional correctness. In: Clarke, E.M., Voronkov, A. (eds.) LPAR 2010. LNCS (LNAI), vol. 6355, pp. 348–370. Springer, Heidelberg (2010). https://doi.org/10.1007/978-3-642-17511-4_20
6. Leino, K.R.M.: Program Proofs. Lulu (2020)
7. Leino, K.R.M., Wüstholz, V.: The Dafny integrated development environment. In: F-IDE. EPTCS, vol. 149, pp. 3–15 (2014)
8. Morgan, C.: Programming from Specifications, 2nd edn. Prentice Hall International (UK) Ltd., Hertfordshire (1994)
9. Morgan, C.: (In-)formal methods: the lost art - a users' manual. In: Liu, Z., Zhang, Z. (eds.) SETSS 2014. LNCS, vol. 9506, pp. 1–79. Springer, Cham (2016). https://doi.org/10.1007/978-3-319-29628-9_1

Teaching Correctness-by-Construction and Post-hoc Verification – The Online Experience

Tobias Runge[1]([✉]), Tabea Bordis[1]([✉]), Thomas Thüm[2]([✉]), and Ina Schaefer[1]([✉])

[1] TU Braunschweig, Brunswick, Germany
{tobias.runge,t.bordis,i.schaefer}@tu-bs.de
[2] University of Ulm, Ulm, Germany
thomas.thuem@uni-ulm.de

Abstract. Correctness of software is an important concern in many safety-critical areas like aviation and the automotive industry. In order to have skilled developers, teaching formal methods is crucial. In our software quality course, we teach students two techniques for correct software development, post-hoc verification and correctness-by-construction. Due to Covid, the last course was held online. We present our lessons learned of adapting the course to an online format on the basis of two user studies; one user study held in person in 2019 and one online user study held after the online course. For good online teaching, we suggest the use of accessible (web-)tools for active participation of the students to compensate the advantages of teaching in person.

1 Introduction

Development of correct software is a concern which is becoming increasingly important. In areas like aviation and the automotive industry, where human lives depend on it, software safety requirements are considerably higher. Therefore, formal methods should be taught to young software developers, so that they learn a reasonable approach to develop correct programs, instead of hacking programs into correctness. With this skill of correct software development, developers can avert major errors in software projects by specifying and verifying the safety-critical parts [35]. Additionally, a development process that includes verification can reduce overall development time because most software is correct from the start, which reduces maintenance time and effort. Besides the prevalent post-hoc verification (PhV) approach, where software is verified after implementation, correctness-by-construction (CbC) [23] is an approach where software is incrementally refined from a specification. In CbC, each refinement step guarantees the correctness of the whole programs. CbC expands the repertoire of programmers with a formal reasoning style that prevents errors in the first place.

Teaching formal methods, the correct specification and verification of programs, is the topic of the Master course Software Quality 2 at TU Braunschweig,

© Springer Nature Switzerland AG 2021
J. F. Ferreira et al. (Eds.): FMTea 2021, LNCS 13122, pp. 101–116, 2021.
https://doi.org/10.1007/978-3-030-91550-6_8

Germany. In the course, students learn the basics of deductive software verification and correctness-by-construction on the practical example of specifying and verifying Java code. In the corresponding exercises, the students solve tasks using corresponding tools. In contrast to previous years, we offered the course of last term online (due to the pandemic).

A difficulty in teaching formal methods is that courses are based on a lot of formal background which may discourage students. Therefore, we highlight the benefit to integrate practical experiences of practical tool usage in teaching that help students to consolidate the taught topics [17]. With tools, students receive immediately feedback if their found solutions are correct. They can also work on larger tasks that are not doable on pen-and-paper. A problem here is the effort to install various tools on different machines, especially when students use their own machines due to online teaching. Tools should be easy to install or web-based such that students enjoy active participation in lectures and exercises of formal method courses. Some good examples for tool support are KeY [3], Whiley [30], and Dafny [25].

Besides an experience report on our online course, we evaluate the learning success of the students with two user studies. We compare the results of an online user study with an earlier user study conducted in the same course, but in person [33]. In the qualitative user study, we evaluate how students solved tasks with two verification techniques post-hoc verification and correctness-by-construction. They used the tools KeY [3] (as instance of a PhV tool) and (Web)CorC [32] (as instance of a CbC tool). Therefore, our user study has four quadrants. We compare CbC with PhV and online with in person courses. With the data from these two studies, we share our lessons learned in transferring the course to an online format, we discuss the quality of the online course, and point out challenges and opportunities for improving online courses in the future. We also compare the web version WebCorC with the previous version CorC to determine important aspects of good tool support. Furthermore, we compare how well students interact with CbC and PhV by collecting their user feedback.

As a result, we confirmed the findings of the first user study. The students made fewer defects in the code with PhV than with CbC, but overall the results are worse than in the first user study. This indicates a worse learning outcome due to the online format. The qualitative questionnaire was answered in favor of CbC. The students liked the structured reasoning of CbC and rated the support provided by the CorC tool as more suitable for finding defects than KeY for PhV. For online teaching, we detect that easily accessible tool support is beneficial for participation in exercises. Additionally, courses on formal methods should be interactive to encourage student participation, thus we discuss how to improve online teaching.

2 Related Work

Teaching formal methods has been discussed by many researchers [11,14,17, 26]. They discuss their teaching experiences and evaluate the learning success of students with respect to different tools and teaching strategies. In detail, Liu et al. [26] highlight the benefits of a good mix of pen-and-paper and tool supported exercises. On paper, students consolidate what they learned without being supported by a tool, and with tools they increase their productivity and learn to analyze defects in the specifications or programs. The interest of students also rises if they can solve exercises on tools and get positive feedback by verifying programs. Creuse et al. [14] mention that teaching by example is beneficial for an easier and more practical entry into formal methods. That the students demand immediate and good feedback on their specification or verification process and want to understand clearly occurring problems is identified by Catano [11]. We differ from this related work [11,14,17,26] that does not examine the aspects of online teaching. In this paper, we discuss new challenges regarding online teaching by comparing our course with a previous course held in person.

With respect to the user study, we compare it with related work that evaluated the usage of verification tools. Petiot et al. [31] evaluated the interaction of programmers with verification tools. The authors analyzed how programmers can be supported when they encounter an open proof goal. To improve user feedback, they categorize defects and calculate counter examples. In the work of Johnson et al. [22], developers were interviewed about their usage of static analysis tools. They also recognized that developers need good error reporting. The influence of formal methods in code reviews was studied by Hentschel et al. [20]. In their study, the symbolic execution debugger (SED) had a positive impact on the location of defect in programs. KeY was also evaluated to analyze how participants interact with the tool during the verification process [9,10]. Back [5] evaluated in an experiment that good tool support is necessary to develop correct software with a refinement based approach. Additionally, he discovered an iterative procedure to refine an incomplete or partially incorrect invariant to a final solution, an insight that we confirm in our first user study [33]. In our user study, we focus more on the usability of the tools during program constructions, how programmers utilize the tools and adapt their programming procedure. In our study, we used KeY as automatic verifier. Thus, we excluded the expertise on interactive proving from our study. We focused on the development of correct programs guided by a specification. In our study, the participants have to implement the programs by themselves.

3   Teaching Formal Methods – Software Quality 2

In this section, we describe the structure of the formal methods course Software Quality 2 at TU Braunschweig. We compare the previous course held in person with the current course in an online format. The goal of this course is to teach students deductive post-hoc verification and correctness-by-construction such that students are able to construct correct programs with these approaches. The course is named Software Quality 2, as we also offer a course Software Quality 1 that focuses on testing software.

In Person Course. The course in person is divided into two parts, 12 lectures with 11 corresponding exercises, each one lasting 90 min. The students attending this course are mostly Master students that had at least two courses in programming, and three courses in theoretical computer science. Normally, between 20 and 50 students attend the course. The lectures are a presentation on topics like design by contract, software model checking, sequent calculus, deductive verification, specifying programs with JML, verifying programs with KeY, and constructing correct programs with CbC. The presentations include some intermediate questions to the audience (e.g., to complete examples) and questions in the end to consolidate the lessons learned. We provide a video of each lecture, so students can prepare the exams by rewatching specific lessons. In the videos, the slides and the audio of the lecture are recorded.

The exercises are divided between pen-and-paper exercises for writing first-order logic and using the sequent calculus, and tool-supported exercises. For the tool-supported exercises, we use different tools: OpenJML [13] to show testing with JML annotated code, Java Pathfinder [19] for software model checking, KeY [3] for program verification, and CorC [32] for correct-by-construction software development. All these tools are pre-installed on machines at the university such that exercises are performed smoothly. The exercises are mostly interactive such that the students solve the tasks and present the solutions, and these solutions are discussed with the audience. This structure of the exercises should consolidate knowledge better than a frontal presentation of solutions.

The course exam is oral. We have a small group of students such that oral exams are feasible. In these exams, we can check whether students understood the topics of this course and can answer cross-cutting questions, and whether they can apply learned techniques to solve code specification and verification tasks. Oral exams are more time-consuming than written exams, but as a teacher one gets better feedback on whether students have understood the content.

Online Course. The setting for the online course is the same as for the previous courses in person: 12 lectures with 11 exercises with weekly meetings covering the same topics. The number of students slightly increased to around 60 students. We upload videos of the recorded lectures of the previous year. Additionally, we give a short recap of the topic followed by a discussion where students can ask questions in the weekly meeting. For the exercises, we upload exercise sheets with tasks that have to be prepared as homework. If the task includes the use of tools, we add an instruction for the installation and usage. In the weekly online session for the exercises, the students are asked to present their solutions which are discussed with the audience afterwards. Thereby, we use Google Docs documents that can be edited by everyone in the session to collect and store the correct answers for exam preparation.

To keep the interactive character of the exercise in the online course, we use the same tools in the exercises as we do for the course in person (i.e. OpenJML, Java Pathfinder, KeY, and CorC). Due to the format of an online session, we could not monitor whether students were actually actively participating. For the tools, we tried to find the easiest way with as few steps as possible for the installation. However, with most of the tools we had some problems with the installation due to outdated documentations or the need of specific JDK or Eclipse versions such that finding a good solution was time-consuming.

The oral exams are held online in the video conference system provided by our university. The student has to attend with camera such that we can check that the right person is taking the exam and that no other persons are in their room. An advantage of taking the oral exam online is that we are able to include practical tasks using tools introduced in the exercises. To omit difficulties in the installation, we installed the tools on our computer and shared the screen. The students then have to explain what they would do and what result they expect.

4   Verification Techniques and Tool Support

We compare in our user study, how students solve tasks using post-hoc verification (PhV) and correctness-by-construction (CbC). We briefly introduce PhV and CbC in the following.

4.1   Post-hoc Verification

The post-hoc verification process [3], verifies the correctness of programs after the implementation. A prover checks that the implementation complies with the pre-/postcondition specification. PhV does not give a development guideline, such that programmers can freely implement the programs as long as the specification is in the end fulfilled. This free process decreases the time to construct a first (potentially faulty) version of a program, but can increase the time to construct a verified version, as it is more likely that defects occur in the code [35].

As an instance, KeY [3] verifies the correctness of Java programs that are specified with the Java modelling language (JML). Starting from a specified program, KeY symbolically executes the programs and closes the remaining proof obligations (semi-)automatically. As we are focusing on the programming and specification aspects in our user study, we use KeY as an automatic tool. This goes along with most programmers not having a theoretical background to verify programs interactively.

Besides KeY, there are a number of tools in the area of specification and program verification: the language Eiffel [27] with the verifier AutoProof [34], the languages SPARK [8], Dafny [25], and Whiley [30], and the tools OpenJML [13], Frama-C [15], VCC [12], VeriFast [21], and VerCors [4]. All languages and tools are candidates to be compared with the CbC methodology, but we decided for KeY because of the previous familiarity of our study participants. Since we used only a subset of the Java language without method calls or custom objects, the difference to other programming languages is minimal.

4.2   Correctness-by-Construction

Correctness-by-construction [16,23,28] is a methodology to incrementally construct correct programs. Starting with a pre-/postcondition specification and an initially abstract program, refinement rules are applied to create an implementation that fulfills the specification. The correctness is guaranteed by the rules if specific side conditions for their applicability hold. Dijsktra [16] and Kourie and Watson [23] identified that the CbC process guides programmers to a correct implementation that has low defects rates and is of better structure than a program ad hoc hacked to correctness. A disadvantage of CbC is the fine-grained refinement process that programmers must adhere to. This complicates program construction for inexperienced programmers, but the fine-grained development with the explicit specification in each node raises awareness for defects in the mind of the programmer [35].

Besides the CbC approach proposed by Kourie and Watson [23], there are other refinement based approaches that guarantee the correctness of the program under development. In the Event-B framework [1], specified automata-based system are refined to concrete implementations. It is implemented in the Rodin platform [2]. In comparison to the CbC approach used here, the abstraction level is different. CbC uses specified source code instead of automata as main artifact. Morgan [28] and Back [7] proposed also related CbC approaches. Morgan's refinement calculus is implemented in the tool ArcAngel [29]. Back et al. [5,6] developed the tool SOCOS. In comparison to CbC, SOCOS starts with invariants additionally to a pre-/postcondition specification.

The tool CorC [32] implements the CbC process in a graphical and textual editor. CorC supports developers by offering refinement rules as proposed by Kourie and Watson [23] to develop programs and checking the correctness of each applied refinement with a program verifier KeY [3]. In CorC, a programmer builds stepwise a correct method by getting feedback directly when one refinement is not correct, for example, when the programmer specifies an invariant that is not satisfied at the beginning of the loop.

WebCorC[1] is an adaption of CorC [32]. Similar to CorC, we decided for the graphical editor in WebCorC because of the student feedback collected during the Software Quality courses. The graphical editor helps students learn the CbC approach by visualizing all important aspects of specifying and refining a program into a correct result. CorC

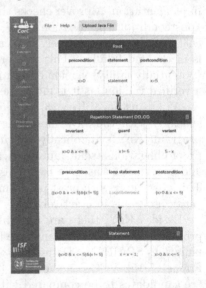

Fig. 1. Program construction in WebCorC

is implemented in Java in the Eclipse framework. In WebCorC, we transferred the graphical editor using a client-server structure, reusing the logic of CorC on server side, but redeveloping the graphical editor as web-frontend. In comparison to CorC, the implementation of WebCorC has no detailed feedback in a console

---

[1] https://www.isf.cs.tu-bs.de/WebCorC/.

when a refinement cannot be proven. This feedback was added only after the user study.

In Fig. 1, we show a program construction in WebCorC. At the top, we specify in the first gray node the program under development. The precondition states that an integer x is greater than zero. In the postcondition, x should be equal to 5. We solve this problem by introducing a loop statement in the first refinement step, called repetition in WebCorC. We introduce a repetition statement for illustration purposes. Of course, an assignment directly solves the problem. For the repetition, we need additional specification: a loop guard, a loop invariant, and a variant. We continue the loop as long as x is not equal to 5. The loop body introduced in the next refinement step, the third node at the bottom, increments x by one. Both refinement steps are checked by WebCorC to be correct.

5   User Study Design

In this section, we describe the design of our user study that was conducted online after the end of the Software Quality 2 course. The goal of this evaluation is to compare the results of the students with the results of a previous study that was held in person. Therefore, we adopt the design of the previous study [33]. We want to get insights into the learning success of the students whether the online course and the use of WebCorC leads to noticeable differences in the outcome. To better compare both studies, we explain the commonalities and differences in the user study design in the following. Note that we used CorC in the first user study and WebCorC in the second user study. If we talk about both tools, we write (Web)CorC.

5.1   General User Study Design

The user study is designed such that students solve two programming tasks each with a different tool. We compare correctness-by-construction and post-hoc verification with the tools (Web)CorC and KeY. Starting with a pre-/postcondition specification, an algorithm should be implemented and verified.

Objective. We want to evaluate whether the CbC approach has a positive or a negative impact on the programming results. We consider the following two research questions to evaluate the verification approaches and tools.

RQ1: What kind of errors do participants make with CbC and PhV?
RQ2: To which degree do participants prefer CbC over PhV?

To evaluate the usability of CbC and PhV, we take the user experience questionnaire (UEQ), and ask the questions OQ1−OQ8:

OQ1: How do you rate your overall work with WebCorC from 1 (very bad) to 5 (very good)?
OQ2: What is your general process when solving tasks with WebCorC/KeY?

OQ3: Do you prefer a web-frontend over the Eclipse environment and why?
OQ4: Were there any specific obstacles during the task execution process?
OQ5: Is the construction of a program by modeling through a refinement structure helpful and why?
OQ6: Do you prefer WebCorC or KeY in general and why?
OQ7: Which of these two tools would you use for verification and why?
OQ8: Which tool better supports avoiding or fixing defects and why?

The UEQ [24] is a standardized test to measure six usability properties of a tool. A participant is asked to rate the tool with 26 items. Each item is a pair of adjectives that describe the tool, one negative and one positive adjective. The user can rate on a 7-point Likert-scale which of the adjectives and to what extent fits more. The range for the answers is between $+3/-3$.

General Design Decisions. The user study is limited to 90 min. To compare both tools, each participant should implement an algorithm with each tool. We set 30 min per task. Thus, the algorithms should be implementable and verifiable in this time frame. We decide for algorithms with a size of under ten lines, but including a loop to have participants writing loop invariants. We give the pre-/postcondition specification of the algorithms so that all participants have the same starting point. This reduces the divergence and lead to comparable programming results. Writing the pre-/postcondition would also cost too much time in this experiment. We decide for the tools (Web)CorC and KeY because the participants have experience with these tools which increases the expressiveness of this study. The material of the user study is published on GitHub.[2]

Tasks. Two algorithms must be correctly implemented and verified. The algorithm `minimum element` calculates the index of the minimum element in an array. The array is non-empty to omit the special case from the algorithm. The algorithm `modulo` calculates the remainder from two input integers; a dividend a and a divisor b. In the algorithm, division and modulo operators are prohibited. The algorithms are similar in size and cyclomatic complexity.

We design the tasks to be small enough to be doable in the time frame. We also design them such that both (Web)CorC and KeY can be used to implement them correctly. For both tasks, assignments, conditional statements and loops where sufficient.

The groups of participants are arranged with the Latin square design. Group A uses (Web)CorC for the first task, and PhV afterwards. Group B does the same tasks but the tools in different order. The tools are switched to address the possibility of learning effects by forcing a specific tool order.

Participants. The participants are students at TU Braunschweig, Germany attending the Software Quality course. These students were taught the fundamentals of software verification, and they learned to use the tools (Web)CorC and KeY. During the course they implement, specify, and verify methods with both tools. We analyzed the programming experience of the participants with a questionnaire [18]. To weaken the restraint against a group of students, we had

---

[2] https://github.com/Runge93/UserstudyCbCPhV.

several students with two to five years of programming experience in industry. Therefore, the participants can be compared with junior developers. The students freely attended the user study. We told them the goal of this experiment, and we offered a monetary payment for attendance. We raffled two times € 25.

Variables. We have the tools as independent variable in our user study, with the treatments CbC and PhV. The correctness of the task were checked with KeY using the automatic mode. For CbC, we checked the task by reverifying each refinement step. When some task was not verifiable, we manually checked for defects in the program or specification. These defects were counted by line.

5.2   Differences in the First and Second User Study

In the first user study, we have 10 participants in two groups who have no significant difference in the programming experience [33]. In the second user studies, we have 13 participants. With a programming experience questionnaire [18], we measure a similar experience in both groups. A value of 2.137 for group A and 2.550 for group B^3. With a Mann-Whitney test, no significant difference between the two groups is measured.

In the first user study, we prepare machines at the university with CorC and KeY. They directly implement both tasks in the Eclipse IDE. Here, they can interact with KeY directly to get feedback about the verification status. In the second user study, we prepare a workspace where the participants can develop one of the algorithms with WebCorC. For the PhV process, they can use their preferred IDE. When they want to verify the algorithm, they upload it and get feedback about the success of the verification. This process can be repeated until a verified result is achieved. As the participants in the second study only upload files in the post-hoc verification tasks and do not interact with KeY directly, we abandon the UEQ for the tool KeY.

In the first user study, we monitored all participants in a controlled experiment in person. In the second user study, we adapt this by having an audio conference. Due to legal restrictions, we cannot use proctoring tools.

6   Results and Discussion

In this section, we show the results of our user study. We evaluate the implementations of each participant by looking at the final result. We focus on a qualitative evaluation of the programming procedure and results of CbC and PhV. Additionally, we evaluate the answers of our questionnaire (UEQ and OQ1–OQ8) to complete the discussion.

6.1   Defects in Implementation and Specification

To answer the first research question, we analyze defects in the code and the specification. By code, we refer to the implemented algorithm without the specification. By specification, we refer to auxiliary annotations such as loop invariants.

---

^{3} The calculation is explained in the work by Feigenspan et al. [18].

Table 1. Defects in code and specification of the final programs of participants

#Defects	PhV 1st		CbC 1st		PhV 2nd		CbC 2nd	
	Code	Spec.	Code	Spec.	Code	Spec.	Code	Spec.
No defects	8	2	4	3	9	1	2	1
Minor defects	1	7	3	4	4	10	4	5
Major defects	1	0	1	0	0	0	0	0
Incomplete	0	1	2	3	0	2	7	7

We classify a program to have major defects, if the program cannot be corrected without rewriting the algorithm. Otherwise, we classify it to have minor defects. The same classification applies for defects in the specification.

Table 1 shows the defects the participants have in their final result when they finished a task in the first and second user study. When we compare CbC and PhV, the participants have fewer coding defects with PhV than with CbC in both user studies. The coding results for PhV and CbC are generally better in the first user study, we have more results without defects for both approaches. Across all participants, a typical defect is a loop guard with a wrong logical comparison operator. With CbC, a recurring problem is that participants forget to initialize variables correctly. In both studies participants have incomplete results in the program for CbC. In the second study, seven participants have not completed the program for CbC.

When we compare defects in the auxiliary specification, more participants have no defects with CbC in the first user study compared with the results for PhV. In the second user study, for each approach only one participant has no defects. In general, the specification results are better in the first user study. More participants have no defects with PhV and CbC. Typical defects with PhV are a missing variant or missing checks whether variables stay in a specific boundary (e.g., out of bounds checks in arrays). With CbC, a common specification defect is that the invariant does not hold initially or after the last loop iteration. However, the participants do not forget the variant when they specify a loop. In both user studies and with both approaches, we have incomplete specifications. Five incomplete results of the auxiliary specification are due to incomplete code.

6.2 User Experience

For the evaluation of the user experience, we show the results of the user experience questionnaire in Fig. 2. The blue results for CorC are from the first user study [33], the red results for WebCorC are from the second study. The answers of the participants are combined into six measurements: *attractiveness, perspicuity, efficiency, dependability, stimulation,* and *novelty*. Except for *efficiency*, the results are better in the first user study. The largest differences are in the scales of *stimulation* and *dependability*. The *stimulation* is rated lower because some participants rate WebCorC *demotivating*. Participants also rate WebCorC as

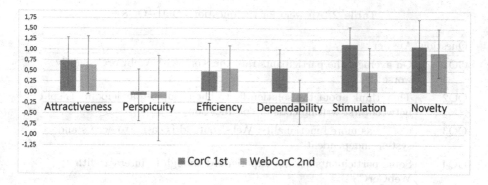

Fig. 2. Results of the user experience questionnaire

unpredictable which results in a negative score for *dependability*. The items *easy to learn/difficult to learn* and *complicated/easy* are answered differently resulting in a big variance for the *perspicuity* measurement in the second user study.

For question OQ1-OQ8, common answers of the participants are summarized in Table 2. Some participants dislike the limited feedback of WebCorC in comparison to CorC, but they prefer the web-frontend due to the easy accessibility. The general process of solving tasks is split between writing specification or the program first. When comparing WebCorC with KeY, the majority of participants prefer WebCorC to solve verification tasks because of the structured process. The participants in favor of KeY argue that they are more familiar with textual programming.

6.3   Discussion of the Research Questions

RQ1. When we compare the defects in code, the participants have similar defects in both approaches (e.g., incorrect loop guards or incomplete invariants), but they have fewer defects with PhV. A possible reason is the familiar environment of writing Java code in a textual editor. Overall, we have worse results in the second study. Regarding the complete results, we explain the difference between both studies with the better feedback of CorC in comparison to WebCorC such that students can find defects more easily. Another reason is that we monitored active participation in the exercises in person. For the online course, we cannot confirm this. It seems that the students were better prepared in the first user study. We noticed that considerably more participants in the second user study have not the necessary knowledge to construct programs with CbC. Some students may not have participated in the exercises and may not have familiarized themselves with (Web)CorC.

RQ2. We answer the second research question, whether participants prefer CbC or PhV. Participants like the familiar programming style with PhV, but the majority prefer (Web)CorC over KeY. The participants mention that CorC has better and fine-grained console feedback which helps detecting defects during program construction. In the previous study, the participants highlight the good

Table 2. Answers for the questions OQ1–OQ8

Question	Answer
OQ1	On average, the participants rate the work with WebCorC slightly worse (2.1/5)
OQ2	They think about the solution first. The group of participants is split between first writing code or specification
OQ3	CorC has more functionality. Web-frontend is easier to access and system independent
OQ4	Some participants are not experienced enough to interact with WebCorC
OQ5	Participants find defects in corner cases with WebCorC. They divide the problem into smaller blocks. CbC rules are too restrictive. Some are unfamiliar with graphical programming
OQ6	Six answers in favor of (Web)CorC. CorC has better feedback than WebCorC. Two participants prefer KeY because of the familiar programming style
OQ7	Six answers in favor of (Web)CorC. Two answers in favor of KeY
OQ8	Six answers in favor of (Web)CorC, mostly because of the better feedback for verification results. Two answers in favor of KeY, as KeY shows the whole proof tree

feedback for each refinement step, which is not implemented in WebCorC yet. With better feedback, they would prefer WebCorC over CorC due to the easier handling and installation. Surprisingly, nobody complains about the additional specification effort in CbC.

Compared to the first UEQ shown in Fig. 2, we get slightly worse results in the second study. The main reason is worse user feedback for WebCorC in comparison to CorC. This insight coincides with the answers of the open questions in both user studies. Thus, participants rate WebCorC more demotivating, unpredictable, and harder to learn because CorC is more advanced. Due to the online course, it was also harder to teach the tools to the students. Students asked fewer questions, therefore, problems were not discovered that also arose during the user study (e.g. the correct initialization of variables). In person, problems stand out more quickly and can be easily explained. Nevertheless, the participants in both studies prefer (Web)CorC over KeY. Considering that the students have more defects with (Web)CorC, the students seem to factor in that they like the CbC approach for correct software development. The main reason against (Web)CorC is that participants are more familiar with the programming style in KeY. A limitation that is likely due to the shorter time of working with the CbC process.

6.4   Threats to Validity

External Validity. The user studies had 10 and 13 participants. With this limited number of participants, the generalizability of the results is restricted, but we were able to analyze the programming results of each participant in detail. The participants are all Computer Science students that learned verification in the Software Quality 2 course. Therefore, they are not experts in verification, but should be able to solve smaller examples as the ones asked in this study. Through their statements in the programming experience questionnaire, most students can be compared to junior developers. Furthermore, the small algorithms reduce the generalizability for larger algorithmic problems. Regarding the time frame of a course, a longer study was not feasible.

Internal Validity. The motivation of the participants and their effort of solving the tasks could not be monitored due to the online version of this user study. As the time was limited for each task, most solutions were not verified completely. With additional time, it would be possible for more algorithms to be verified.

7 Lessons Learned for Online Teaching

In this section, we conclude the paper by summarizing our lessons learned for online teaching. The first three findings are based on the results in the user studies. The last three also include our experiences from the online course.

Procedure of Software Development. By analyzing the questionnaire and the programming results of the user study, we notice that students are mostly hacking programs into correctness. By teaching the correctness-by-construction approach, we enable students to think of the specification and the corner cases first, before starting to program. This is well-received by the students, but the approach needs time to be adopted.

Accessibility of Tools. In the questionnaire, students highlight that tools should be easy to install for online teaching. If a tool needs many installation steps or has a high potential to fail on some machines, students will not actively participate in exercises. Students that are not able to solve the CbC tasks indicate missing knowledge in (Web)CorC. Also, tools should be freely available such that students are not excluded because they cannot afford the tool. Many tools that we use during our lectures are Eclipse plug-ins. For Eclipse plug-ins, the easiest way to install them is by using an update site. However, for some tools, the update sites are not accessible anymore or only work with specific versions of Eclipse or JDK. That has to be checked before a course.

Feedback of Tools. From the questionnaire, we know that tools should give detailed and fine-grained feedback if errors occur in the development process. Without feedback, the finding of defects during new tasks gets frustrating such that students tend to give up faster online. This confirms results in the literature [11, 26].

Besides of the user studies, we also collect feedback during the courses. Good teaching is characterized by active participation of students [17, 26]. As students are more quickly distracted online, we describe how to improve the online course such that students actively participate. Regarding the fact that we have more programming defects in the second user study, we still have to improve the online course to be as good as the course in person.

Breakout Rooms. During the online exercises, we found that including tasks where students can work in small groups in breakout rooms increases the number of actively participating students. This holds, especially when the teacher is not constantly in the same room and the students can work together on a task, which has not been prepared in advance. Generally, breakout rooms also help students to connect with each other and build learning groups for exams which became more difficult during the pandemic.

Interactions in an Online Setting. In online courses, it is way more important that students are willing to follow the lecture and to participate in exercises. In the results of the user study, we encounter several students that indicate missing background knowledge for the tasks. To prevent this, we derive the following best practices for online teaching: Students should attend exercises with cameras which increases attention. Students should be integrated into lectures by asking questions. When videos and slides are provided in addition to a lecture, students can consolidate what has been learned. Exercises with voluntary tasks are working only for a minority of students. Other students will attend the exercises unprepared. So exercises need to be mandatory or could provide bonus points for the final exam.

Openness to Novel Approaches. Students are open minded for new techniques and tools as we can see from our experiences with (Web)CorC. As teachers, we have to ensure that new topics are introduced interactively and with examples. However, when the new technique is not introduced properly, students will not consider it for future tasks and fall back to old familiar approaches. We want to ensure that formal methods are not taught for the sake of the course, but be anchored in the mind of young computer scientists. So the introduction of the new techniques needs to be thorough, well illustrated using meaningful examples, and supported by accessible tools.

Acknowledgments. We thank Huu Cuong Nguyen and Malena Horstmann for their help in preparing and conducting the user study.

References

1. Abrial, J.R.: Modeling in Event-B: System and Software Engineering. Cambridge University Press, Cambridge (2010)
2. Abrial, J.R., Butler, M., Hallerstede, S., Hoang, T.S., Mehta, F., Voisin, L.: Rodin: an open toolset for modelling and reasoning in Event-B. STTT **12**(6), 447–466 (2010)
3. Ahrendt, W., Beckert, B., Bubel, R., Hähnle, R., Schmitt, P.H., Ulbrich, M.: Deductive Software Verification-The KeY Book: From Theory to Practice, vol. 10001. Springer, Heidelberg (2016). https://doi.org/10.1007/978-3-319-49812-6

4. Amighi, A., Blom, S., Darabi, S., Huisman, M., Mostowski, W., Zaharieva-Stojanovski, M.: Verification of concurrent systems with VerCors. In: Bernardo, M., Damiani, F., Hähnle, R., Johnsen, E.B., Schaefer, I. (eds.) SFM 2014. LNCS, vol. 8483, pp. 172–216. Springer, Cham (2014). https://doi.org/10.1007/978-3-319-07317-0_5

5. Back, R.J.: Invariant based programming: basic approach and teaching experiences. FAOC **21**(3), 227–244 (2009)

6. Back, R.-J., Eriksson, J., Myreen, M.: Testing and verifying invariant based programs in the SOCOS environment. In: Gurevich, Y., Meyer, B. (eds.) TAP 2007. LNCS, vol. 4454, pp. 61–78. Springer, Heidelberg (2007). https://doi.org/10.1007/978-3-540-73770-4_4

7. Back, R.J., Wright, J.: Refinement Calculus: A Systematic Introduction. Springer, Heidelberg (2012)

8. Barnes, J.G.P.: High Integrity Software: The Spark Approach to Safety and Security. Pearson Education (2003)

9. Beckert, B., Grebing, S., Böhl, F.: A usability evaluation of interactive theorem provers using focus groups. In: Canal, C., Idani, A. (eds.) SEFM 2014. LNCS, vol. 8938, pp. 3–19. Springer, Cham (2015). https://doi.org/10.1007/978-3-319-15201-1_1

10. Beckert, B., Grebing, S., Böhl, F.: How to put usability into focus: using focus groups to evaluate the usability of interactive theorem provers. EPTCS **167**, 4–13 (2014)

11. Cataño, N.: Teaching formal methods: lessons learnt from using Event-B. In: Dongol, B., Petre, L., Smith, G. (eds.) FMTea 2019. LNCS, vol. 11758, pp. 212–227. Springer, Cham (2019). https://doi.org/10.1007/978-3-030-32441-4_14

12. Cohen, E., et al.: VCC: a practical system for verifying concurrent C. In: Berghofer, S., Nipkow, T., Urban, C., Wenzel, M. (eds.) TPHOLs 2009. LNCS, vol. 5674, pp. 23–42. Springer, Heidelberg (2009). https://doi.org/10.1007/978-3-642-03359-9_2

13. Cok, D.R.: OpenJML: JML for Java 7 by extending OpenJDK. In: Bobaru, M., Havelund, K., Holzmann, G.J., Joshi, R. (eds.) NFM 2011. LNCS, vol. 6617, pp. 472–479. Springer, Heidelberg (2011). https://doi.org/10.1007/978-3-642-20398-5_35

14. Creuse, L., Dross, C., Garion, C., Hugues, J., Huguet, J.: Teaching deductive verification through FRAMA-C and SPARK for non computer scientists. In: Dongol, B., Petre, L., Smith, G. (eds.) FMTea 2019. LNCS, vol. 11758, pp. 23–36. Springer, Cham (2019). https://doi.org/10.1007/978-3-030-32441-4_2

15. Cuoq, P., Kirchner, F., Kosmatov, N., Prevosto, V., Signoles, J., Yakobowski, B.: Frama-C. In: Eleftherakis, G., Hinchey, M., Holcombe, M. (eds.) SEFM 2012. LNCS, vol. 7504, pp. 233–247. Springer, Heidelberg (2012). https://doi.org/10.1007/978-3-642-33826-7_16

16. Dijkstra, E.W.: A Discipline of Programming. Prentice Hall, Hoboken (1976)

17. Divasón, J., Romero, A.: Using Krakatoa for teaching formal verification of Java programs. In: Dongol, B., Petre, L., Smith, G. (eds.) FMTea 2019. LNCS, vol. 11758, pp. 37–51. Springer, Cham (2019). https://doi.org/10.1007/978-3-030-32441-4_3

18. Feigenspan, J., Kästner, C., Liebig, J., Apel, S., Hanenberg, S.: Measuring programming experience. In: ICPC, pp. 73–82. IEEE (2012)

19. Havelund, K., Pressburger, T.: Model checking Java programs using Java pathfinder. STTT **2**(4), 366–381 (2000)

20. Hentschel, M., Hähnle, R., Bubel, R.: Can formal methods improve the efficiency of code reviews? In: Ábrahám, E., Huisman, M. (eds.) IFM 2016. LNCS, vol. 9681, pp. 3–19. Springer, Cham (2016). https://doi.org/10.1007/978-3-319-33693-0_1

21. Jacobs, B., Smans, J., Piessens, F.: A quick tour of the VeriFast program verifier. In: Ueda, K. (ed.) APLAS 2010. LNCS, vol. 6461, pp. 304–311. Springer, Heidelberg (2010). https://doi.org/10.1007/978-3-642-17164-2_21

22. Johnson, B., Song, Y., Murphy-Hill, E., Bowdidge, R.: Why don't software developers use static analysis tools to find bugs? In: ICSE, pp. 672–681. IEEE Press (2013)

23. Kourie, D.G., Watson, B.W.: The Correctness-by-Construction Approach to Programming. Springer, Heidelberg (2012)

24. Laugwitz, B., Held, T., Schrepp, M.: Construction and evaluation of a user experience questionnaire. In: Holzinger, A. (ed.) USAB 2008. LNCS, vol. 5298, pp. 63–76. Springer, Heidelberg (2008). https://doi.org/10.1007/978-3-540-89350-9_6

25. Leino, K.R.M.: Dafny: an automatic program verifier for functional correctness. In: Clarke, E.M., Voronkov, A. (eds.) LPAR 2010. LNCS (LNAI), vol. 6355, pp. 348–370. Springer, Heidelberg (2010). https://doi.org/10.1007/978-3-642-17511-4_20

26. Liu, S., Takahashi, K., Hayashi, T., Nakayama, T.: Teaching formal methods in the context of software engineering. ACM SIGCSE Bull. 41(2), 17–23 (2009)

27. Meyer, B.: Eiffel: a language and environment for software engineering. JSS 8(3), 199–246 (1988)

28. Morgan, C.: Programming from Specifications, 2nd edn. Prentice Hall, Hoboken (1994)

29. Oliveira, M.V.M., Cavalcanti, A., Woodcock, J.: ArcAngel: a tactic language for refinement. FAOC 15(1), 28–47 (2003)

30. Pearce, D.J., Groves, L.: Whiley: a platform for research in software verification. In: Erwig, M., Paige, R.F., Van Wyk, E. (eds.) SLE 2013. LNCS, vol. 8225, pp. 238–248. Springer, Cham (2013). https://doi.org/10.1007/978-3-319-02654-1_13

31. Petiot, G., Kosmatov, N., Botella, B., Giorgetti, A., Julliand, J.: Your proof fails? Testing helps to find the reason. In: Aichernig, B.K.K., Furia, C.A.A. (eds.) TAP 2016. LNCS, vol. 9762, pp. 130–150. Springer, Cham (2016). https://doi.org/10.1007/978-3-319-41135-4_8

32. Runge, T., Schaefer, I., Cleophas, L., Thüm, T., Kourie, D., Watson, B.W.: Tool support for correctness-by-construction. In: Hähnle, R., van der Aalst, W. (eds.) FASE 2019. LNCS, vol. 11424, pp. 25–42. Springer, Cham (2019). https://doi.org/10.1007/978-3-030-16722-6_2

33. Runge, T., Thüm, T., Cleophas, L., Schaefer, I., Watson, B.W., et al.: Comparing correctness-by-construction with post-hoc verification—a qualitative user study. In: Sekerinski, E. (ed.) FM 2019. LNCS, vol. 12233, pp. 388–405. Springer, Cham (2020). https://doi.org/10.1007/978-3-030-54997-8_25

34. Tschannen, J., Furia, C.A., Nordio, M., Polikarpova, N.: AutoProof: auto-active functional verification of object-oriented programs. In: Baier, C., Tinelli, C. (eds.) TACAS 2015. LNCS, vol. 9035, pp. 566–580. Springer, Heidelberg (2015). https://doi.org/10.1007/978-3-662-46681-0_53

35. Watson, B.W., Kourie, D.G., Schaefer, I., Cleophas, L.: Correctness-by-construction and post-hoc verification: a marriage of convenience? In: Margaria, T., Steffen, B. (eds.) ISoLA 2016. LNCS, vol. 9952, pp. 730–748. Springer, Cham (2016). https://doi.org/10.1007/978-3-319-47166-2_52

Using Isabelle in Two Courses on Logic and Automated Reasoning

Jørgen Villadsen$^{(\boxtimes)}$ ⓘ and Frederik Krogsdal Jacobsen ⓘ

Technical University of Denmark, Kongens Lyngby, Denmark
jovi@dtu.dk

Abstract. We present our experiences teaching two courses on formal methods and detail the contents of the courses and their positioning in the curriculum. The first course is a bachelor course on logical systems and logic programming, with a focus on classical first-order logic and automatic theorem proving. The second course is a master course on automated reasoning, with a focus on classical higher-order logic and interactive theorem proving. The proof assistant Isabelle is used in both courses, using Isabelle/Pure as well as Isabelle/HOL. We describe our online tools to be used with Isabelle/HOL, in particular the Sequent Calculus Verifier (SeCaV) and the Natural Deduction Assistant (NaDeA). We also describe our innovative Students' Proof Assistant which is formally verified in Isabelle/HOL and integrated in Isabelle/jEdit using Isabelle/ML.

Keywords: Logic · Automated reasoning · Proof assistant Isabelle

1 Introduction

We present our experiences teaching two courses on formal methods at the Technical University of Denmark (DTU):

- BSc Course: DTU Course 02156 Logical Systems and Logic Programming
  https://kurser.dtu.dk/course/02156
- MSc Course: DTU Course 02256 Automated Reasoning
  https://kurser.dtu.dk/course/02256

Both courses are taught in English. Figure 1 shows the objectives and content of the two courses. The objectives need to be approved by the study board and are not expected to change much from year to year. The above links also include some official statistics like evaluations and grades but mostly in Danish. Both courses count for 5 ECTS points, which corresponds to approximately 2 h of lectures and 2 h of group exercise sessions per week, plus individual study and assignment work (expected around 9 h per week in total), for 13 weeks (summing up to 140 h with exam preparations).

© Springer Nature Switzerland AG 2021
J. F. Ferreira et al. (Eds.): FMTea 2021, LNCS 13122, pp. 117–132, 2021.
https://doi.org/10.1007/978-3-030-91550-6_9

(a) Objectives and content of the course Logical Systems and Logic Programming.

(b) Objectives and content of the course Automated Reasoning.

General course objectives
The aim of the course is to give the students an introduction to some of the basic declarative formalisms from formal computer science and logic that can be used for describing, analysing and constructing IT systems. It will cover theoretical insight as well as practical skills in relevant high-level programming languages.

General course objectives
Reasoning is the ability to make logical inferences. The aim of the course is to give the students an introduction to automatic and interactive computer systems for reasoning about mathematical theorems as well as properties of IT systems. It will cover theoretical insight as well as practical skills in relevant proof assistants.

Learning objectives
A student who has met the objectives of the course will be able to:

- relate different kinds of proof systems
- construct formal proofs in elementary logics
- exploit selected classical and non-classical logics
- use the backtracking algorithm for simple problem solving
- analyze the effect of a declarative program
- establish a functional design for a given problem, so that the main concepts of the problem are directly traceable in the design
- master logical approaches to programming in terms of defining recursive predicates
- communicate solutions to problems in a clear and precise manner

Learning objectives
A student who has met the objectives of the course will be able to:

- explain the basic concepts introduced in the course
- express mathematical theorems and properties of IT systems formally
- master the natural deduction proof system
- relate first-order logic, higher-order logic and type theory
- construct formal proofs in the procedural style and in the declarative style
- use automatic and interactive computer systems for automated reasoning
- evaluate the trustworthiness of proof assistants and related tools
- communicate solutions to problems in a clear and precise manner

Content
The course covers logic programming (in particular Prolog as a rapid prototyping tool), elementary logics (including propositional and first-order logic), proof systems (deductive systems and/or refutation systems), and problem solving techniques (for instance the backtracking algorithm).

Content
The natural deduction proof system, first-order logic, higher-order logic and type theory. Formal proofs in the procedural style and in the declarative style using automatic and interactive provers. The Isabelle proof assistant in artificial intelligence and computer science.

Fig. 1. Objectives and course content of the two courses.

The first course is on logical systems and logical programming, and is intended for final-year BSc students (over the years interested students have successfully taken it already at the start of the second year of their bachelor). The course has been given more or less in the same format since 2006 with an increasing number of students, currently around 80 students per year.

The second course is on automated reasoning, and is intended for MSc students (interested students have successfully taken it already during the final year of their bachelor). The course was given for the first time in 2020 and has around 40 students per year.

The main changes due to COVID-19 were online lessons using Zoom and online home exams instead of physical at DTU. We did not make any other changes to the courses and we will not elaborate on the COVID-19 situation in the present paper.

Both of the courses use the proof assistant Isabelle [30] to showcase verified proof systems and provers, which we have implemented in Isabelle. This allows us to discuss common proof methods for e.g. soundness and completeness and allows students to experiment with larger proofs without losing track of what is going on. We also use Isabelle for assignments and exam questions concerning these proof systems. This allows students to get immediate feedback from the proof assistant, and allows us to easily check if the submitted proofs are correct. Recent research indicates that quick formative evaluation has a large impact on learning when teaching introductory computer science [16]. We use Isabelle since it is the proof assistant that we know best and because Isabelle is a generic proof assistant which allows us to use both Isabelle/HOL and Isabelle/Pure as detailed in later sections.

The BSc course additionally revolves heavily around the Prolog programming language, on which we spend around half of the time. Students thus learn to couple logical programming with logic, and we showcase many interesting programs related to the rest of the course content. The MSc course focuses more on functional programming within the Isabelle proof assistant, and how this can be coupled to formal methods and proofs about programs.

To enable this use of Isabelle and Prolog, we need students to hit the ground running so they can use the implementations of the logical concepts from the beginning. We recall the following quote from Donald Knuth:

> When certain concepts of TEX are introduced informally, general rules will be stated; afterwards you will find that the rules aren't strictly true. In general, the later chapters contain more reliable information than the earlier ones do. The author feels that this technique of deliberate lying will actually make it easier for you to learn the ideas. Once you understand a simple but false rule, it will not be hard to supplement that rule with its exceptions.

For Isabelle and Prolog, we throw the students into the deep end and return later to explain how everything actually fits together. Unification, for example, is treated informally until late in the course where students have the logical background to understand how it works and need the details of it in order to master the resolution calculus for first-order logic [3].

On the other hand, we never cut corners about logic itself. With the proof assistant Isabelle/HOL we can create canonical reference documents for logics and their metatheory. The formal language of Isabelle/HOL, namely higher-order logic, is precise and unambiguous. This means every proof can be mechanically checked, and that it is impossible to cheat and omit any details.

We summarize our main points:

1. We use Isabelle in both courses, including the editor Isabelle/jEdit and the Isabelle/ML facilities.
2. By exploring formally verified proof systems and provers, we use formal methods on the field of formal methods itself.
3. In the advanced course we in addition use Isabelle/Pure, showing the generic Isabelle logical framework and forcing students to manage without the automation of Isabelle/HOL.
4. We rely on group exercise sessions with competent teaching assistants and peer assistance in combination with the Isabelle proof assistant and our own tools.
5. We have individual assignments, as often and as early as possible, with a quick feedback loop from the teaching assistants.

In the next section, we discuss related work. In Sect. 3 we detail the position of our courses within the context of the rest of our computer science and software engineering program. Next we describe the BSc course in Sect. 4, followed by a description of the MSc course in Sect. 5. Finally we describe our overall experiences and ideas for future work in Sect. 6 and conclude in Sect. 7.

2   Related Work

Our two courses are based on a number of tools for teaching logic developed in recent years [10–15, 21, 36–40]. In the present paper we elaborate, for the first time, on the courses and detail our experiences.

We are not aware of any textbooks for teaching logic using the Isabelle proof assistant, but textbooks on formalizing a number of other computer science topics exist, like the book on programming language semantics [24, 29] or functional algorithms [26–28]. These books show that the proof assistant Isabelle/HOL can be used for teaching semantics, algorithms and data structures. There are also impressive books for the proof assistant Coq [33] and the proof assistant Lean [1] but we are not aware of approaches to teaching logic and automated reasoning where the proof systems and provers are formalized in a proof assistant. We envision a textbook around our tools, but are currently relying on a number of unpublished smaller notes and tutorials to teach students how to use them.

Bella [2] presents a teaching methodology for the so-called Inductive Method to verified security protocols and notes the following step:

But the first and foremost step is to convince the learners that they already somewhat used formal methods, although for other applications, for example in the domains of Physics and Mathematics. The argument will convey as few technicalities as possible, in an attempt to promote the general message that formal methods are not extraterrestrial even for students who are not theorists.

We attempt to promote a similar message to the students following our courses.

Harrison [19], Blanchette [5] and Reis [34] discuss aspects of formalizing the metatheory of proof systems and provers. In contrast to our work they do not consider the use of such formalizations as central components and tools in logic and automated reasoning courses.

3   Curricular Overview

The first of our courses is the BSc course meant for final-year students, while our second course is the MSc course. We would like to briefly explain the positioning of our courses within the overall computer science and engineering curriculum at the Technical University of Denmark (DTU). The curriculum at DTU is organized in a half-year semester structure, but after the first year students are free to organize their own study plan and have many electives which can be used for any course offered at the institution.

While the MSc course is intended to be followed after our BSc course, our students have very varied backgrounds because many MSc students have BSc degrees from other institutions. The backgrounds of the students following the BSc course are also varied because the course is followed by many exchange students, BEng students, and General Engineering students. Figure 2 shows the context of our courses in the overall computer science and software engineering program.

Course numbers and ECTS points are as follows for the BSc courses:

- 01017 Discrete Mathematics (5 ECTS)
- 02101 Introductory Programming (5 ECTS)
- 02105 Algorithms and Data Structures 1 (5 ECTS)
- 02110 Algorithms and Data Structures 2 (5 ECTS)
- 02141 Computer Science Modelling (10 ECTS)
- 02156 Logical Systems and Logic Programming (5 ECTS)
- 02157 Functional Programming (5 ECTS)
- 02180 Introduction to Artificial Intelligence (5 ECTS)
- 02450 Introduction to Machine Learning and Data Mining (5 ECTS)

We have here omitted the traditional BSc courses in Computer Engineering and Software Engineering as they play only a minor role in this context.

The MSc courses are organized in study lines which are optional to follow, but nevertheless guide the study planning for the students. Except for a single Innovation in Engineering course we do not have any mandatory MSc courses, though

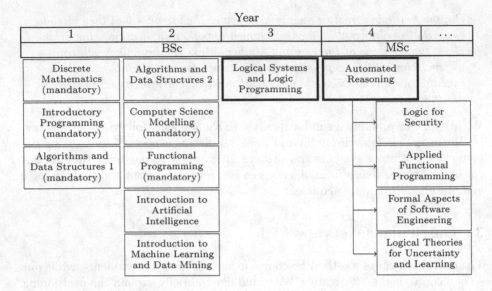

Fig. 2. Suggested course progression surrounding our courses.

students must of course primarily take courses related to computer science. The selected MSc courses to be taken after our courses on logic and automated reasoning are on the following study lines (we also have study lines in Computer Security and Digital Systems, but the former is more practically oriented compared to Safe and Secure by Design and the latter is more Electrical Engineering oriented):

- Study Line: Artificial Intelligence and Algorithms
  02256 Automated Reasoning (5 ECTS)
  02287 Logical Theories for Uncertainty and Learning (5 ECTS)
- Study Line: Embedded and Distributed Systems
  02257 Applied Functional Programming (5 ECTS)
- Study Line: Safe and Secure by Design
  02244 Logic for Security (7.5 ECTS)
- Study Line: Software Engineering
  02263 Formal Aspects of Software Engineering (5 ECTS)

For the BSc course, we recommend that students have previous programming experience as well as knowledge of discrete mathematics and at least basic knowledge of algorithms and data structures. Functional programming is an advantage due to our use of systems implemented in Isabelle/HOL. These prerequisites are obtained in mandatory first and second year courses by most of the students following our BSc course.

However, a significant number of the students following our BSc course are either exchange students, come from the General Engineering program at DTU, or are BEng students. For exchange students, the structure of the curriculum of their home institution may diverge from ours, which means that they sometimes have quite different backgrounds. Students from the General Engineering program have an interdisciplinary study plan, which means that they may not have all of the recommended prerequisites. Finally, BEng students have a curriculum which differs significantly from that of the BSc students, and are generally more focused on practical applications.

For the MSc course, we recommend that students have followed our BSc course and have experience with functional programming and basic algorithms in artificial intelligence. Students coming from the BSc program at DTU will mostly have these prerequisites, but a large amount of students on our MSc programmes come from other institutions. This means that we generally need to assume that students will not have all of the recommended prerequisites, and especially that they have not followed our BSc course.

Our courses provide skills that are useful in a number of MSc courses at DTU. A firm grasp of logic is of course useful for courses such as Logic for Security and Logical Theories for Uncertainty and Learning. Familiarity with formal methods and logic is useful for a course on Formal Aspects of Software Engineering. Several topics covered in our courses can provide interesting project ideas to implement for a course on Applied Functional Programming or for a BSc or MSc thesis. At DTU, it is also quite common to organize special elective courses based on student interest in a specific topic, and we have done so based on advanced topics related to our courses several times.

Both of our courses consist of a mix of lectures, live demonstrations of programs and proofs in Isabelle, and exercise sessions. During exercise sessions, students are free to discuss the problems within groups, and teaching assistants are available to provide help and formative evaluation during the sessions. Since many exercise sessions concern systems implemented in Isabelle, students can get immediate feedback on their proofs, and may ask teaching assistants for more detailed feedback and help if this is not sufficient. To aid student independence, we have for some of our systems developed tools which can provide more detailed formative evaluation of student work than Isabelle. Solutions are provided after all exercise sessions so students can compare their own proofs with ours. This is in contrast to the assignments where only feedback is provided.

Both courses additionally have several individual assignments, which we grade and provide feedback on quickly. These assignments count for part of the overall grade of the courses, with the rest of the grade coming from the exam.

4   BSc Course: Logical Systems and Logic Programming

The first of our courses is the BSc course on Logical Systems and Logic Programming. The course is essentially split in two concurrently running parts. One part of the course covers logic programming in Prolog, while the other part concerns formal logic. The course is based primarily on the textbook Mathematical Logic for Computer Science by Ben-Ari [3], and we cover most of the book in the course.

The course learning objectives can be seen in Fig. 1a and the week-by-week plan of the course can be seen in Table 1.

Table 1. Course plan for the course on Logical Systems and Logic Programming.

Week	Main topics	Assignment
1	Tutorial on Logic Programming	
2	Introduction (Prolog Note)	
3	Propositional Logic: Formulas, Models, Tableaux	
4	Propositional Logic: Deductive Systems	X
5	Propositional Logic: Sequent Calculus Verifier—Isabelle	
6	Propositional Logic: Resolution	X
7	First-Order Logic: Formulas, Models, Tableaux	
8	First-Order Logic: Deductive Systems	X
9	First-Order Logic: Sequent Calculus Verifier—Isabelle	
10	First-Order Logic: Terms and Normal Forms	X
11	First-Order Logic: Resolution	
12	First-Order Logic: Logic Programming	
13	First-Order Logic: Undecidability and Model Theory	X

We start by introducing the basic features of Prolog through a number of examples and exercises. We continue to introduce more Prolog features throughout the course, and use Prolog to show how to implement many of the concepts in logical systems.

After the short introduction to Prolog, we begin covering propositional logic. Following Ben-Ari's book, we cover formulas, semantics, models, and semantic tableaux. This also allows us to discuss the issues of soundness and completeness. Next, we cover deductive systems in the styles of Hilbert and Gentzen, and show how to prove completeness by relating systems to existing systems that are known to be complete.

Having done this, we take an excursion into formal methods by introducing the Isabelle proof assistant. We use our Sequent Calculus Verifier (SeCaV) [11, 12,15], which is implemented in Isabelle/HOL, to teach students how to write and formally verify proofs. This allows students to experiment with their proofs while getting immediate feedback on their correctness. For this first introduction, we use a version of SeCaV which is restricted to propositional logic. Since SeCaV is implemented within Isabelle/HOL, this also exposes students to the basics of proofs in the Isar proof language of Isabelle.

To conclude the sessions on propositional logic we introduce resolution, including Prolog programs that implement each step of a proof by resolution. This allows students to experiment with resolution proofs while also exposing them to non-trivial Prolog programs.

Next, we go through essentially the same topics as before, but now for first-order logic. We again use Prolog programs to explain concepts such as Skolemization and include a Prolog program for resolution in first-order logic.

At this point, we again digress to explore the full version of our Sequent Calculus Verifier, which is a deductive system for first-order logic. The system allows us to explain concepts such as de Bruijn indices and substitution of bound variables with simple implementations. Additionally, we showcase the formal proofs of soundness and completeness for the system. This allows us to explain these proofs in much detail while exposing students to more advanced usage of Isabelle. The implementation of SeCaV in Isabelle/HOL is also a good example for students, since it includes fully elaborated and concrete implementations of e.g., syntax, semantics, and proof rules.

Having done this, we include a number of exercises on implementing logical concepts in Prolog, including the implementation of a SAT solver. We briefly introduce concepts such as higher-order programming and constraint programming in Prolog. We also "close the loop" by finally explaining the relation between logic programming in Prolog and first-order logic. At this point the students have been sufficiently exposed to both to understand this quite quickly.

The final lecture is spent discussing some simple results in model theory and the concept of undecidability.

Throughout the course, students must hand in assignments concerning the various topics of the course. The first assignment is a mix of pen-and-paper formal proofs and Prolog programming exercises, while later assignments also include formal proof exercises in the Sequent Calculus Verifier. These assignments contribute to the final grade of the course. The rest of the grade is determined by a written final exam, which also includes a mix of pen-and-paper formal proofs and Prolog programming exercises.

5 MSc Course: Automated Reasoning

The second of our courses is the MSc course on Automated Reasoning. The course is essentially split in two concurrently running parts. One part of the course covers proving and programming in Isabelle [22,25], while the other part concerns formal logic [10–15,21,36–40]. The course learning objectives can be seen in Fig. 1b and the week-by-week plan of the course can be seen in Table 2.

We start by exploring our formally verified micro provers for propositional logic [37,38], which allow us to explain how provers can be implemented in e.g., Haskell, Isabelle/ML and Standard ML and how to prove correctness in Isabelle.

The Natural Deduction Assistant (NaDeA) [39] is a browser application for classical first-order logic with constants and functions. The syntax, the semantics and the inductive definition of the natural deduction proof system along with the soundness and completeness proofs are verified in Isabelle/HOL. Finished NaDeA proofs are automatically translated into the corresponding Isabelle-embedded proofs.

We have developed teaching materials about Isabelle/Pure [41], showing the generic Isabelle logical framework in order to ensure that students understand

Table 2. Course plan for the course on Automated Reasoning.

Week	Main topics	Assignment
1–2	Prerequisites, micro provers, getting started with Isabelle	X
3–4	Natural Deduction Assistant (NaDeA)	X
5–6	Isabelle/Pure for Intuitionistic and Classical First-Order Logic	X
7–8	Isabelle/Pure for Intuitionistic and Classical Higher-Order Logic	X
9–10	Axiomatic Propositional, First-Order and Higher-Order Logic	X
11–12	Students' Proof Assistant (SPA)	
13	Reserve/buffer lecture	X

what is going on at a lower level when they use the automation of Isabelle/HOL, and the learning outcome is tested in assignments using Isabelle/Pure.

We briefly describe our route from axiomatic propositional logic [7] to first-order logic with equality in our Students' Proof Assistant (SPA) [36] running inside Isabelle/HOL with a formally verified LCF-style prover kernel [31] and declarative proofs [41,42].

The students can experiment in Isabelle/HOL with our formalized soundness and completeness theorems for several axiomatic systems [10], including the following well-known axioms in addition to the rule modus ponens:

Wajsberg 1937
$$p \Rightarrow (q \Rightarrow p)$$
$$(p \Rightarrow q) \Rightarrow ((q \Rightarrow r) \Rightarrow (p \Rightarrow r))$$
$$((p \Rightarrow q) \Rightarrow p) \Rightarrow p$$
$$\bot \Rightarrow p$$

Wajsberg 1939
$$p \Rightarrow (q \Rightarrow p)$$
$$(p \Rightarrow (q \Rightarrow r)) \Rightarrow ((p \Rightarrow q) \Rightarrow (p \Rightarrow r))$$
$$((p \Rightarrow \bot) \Rightarrow \bot) \Rightarrow p$$

Łukasiewicz 1948
$$((p \Rightarrow q) \Rightarrow r) \Rightarrow ((r \Rightarrow p) \Rightarrow (s \Rightarrow p))$$
$$\bot \Rightarrow p$$

We extend the Wajsberg 1939 axiomatic system for propositional logic to first-order logic with equality [20]:

$$\vdash q \text{ if } \vdash p \Rightarrow q \text{ and } \vdash p \quad \text{(modus ponens rule)}$$
$$\vdash \forall x.p \text{ if } \vdash p \quad \text{(generalization rule)}$$

$$\vdash p \Rightarrow (q \Rightarrow p)$$
$$\vdash (p \Rightarrow (q \Rightarrow r)) \Rightarrow ((p \Rightarrow q) \Rightarrow (p \Rightarrow r))$$
$$\vdash ((p \Rightarrow \bot) \Rightarrow \bot) \Rightarrow p$$
$$\vdash (\forall x.p \Rightarrow q) \Rightarrow (\forall x.p) \Rightarrow (\forall x.q)$$
$$\vdash p \Rightarrow (\forall x.p) \text{ provided } x \notin \text{FV}(p)$$

$$\vdash (\exists x.x = t) \text{ provided } x \notin \text{FVT}(t)$$
$$\vdash t = t$$
$$\vdash s_1 = t_1 \Rightarrow \cdots \Rightarrow (s_n = t_n \Rightarrow f(s_1, \ldots, s_n) = f(t_1, \ldots, t_n))$$
$$\vdash s_1 = t_1 \Rightarrow \cdots \Rightarrow (s_n = t_n \Rightarrow P(s_1, \ldots, s_n) = P(t_1, \ldots, t_n))$$
$$\vdash (p \Leftrightarrow q) \Rightarrow (p \Rightarrow q)$$
$$\vdash (p \Leftrightarrow q) \Rightarrow (q \Rightarrow p)$$
$$\vdash (p \Rightarrow q) \Rightarrow ((q \Rightarrow p) \Rightarrow (p \Leftrightarrow q))$$
$$\vdash \top \Leftrightarrow (\bot \Rightarrow \bot)$$
$$\vdash \neg p \Leftrightarrow (p \Rightarrow \bot)$$
$$\vdash (p \wedge q) \Leftrightarrow ((p \Rightarrow (q \Rightarrow \bot)) \Rightarrow \bot)$$
$$\vdash (p \vee q) \Leftrightarrow \neg(\neg p \wedge \neg q)$$
$$\vdash (\exists x.p) \Leftrightarrow \neg(\forall x.\neg p)$$

Here FV is the set of free variables in a formula and FVT is the set of free variables in a term. Note that the axiomatic system is substitutionless as it uses equality in a clever way to avoid the complications of substitution [20,36].

Amongst Pelletier's problems [32] for automated reasoning is problem 34, which is also known as Andrews's Challenge. The proof is not obvious at first glance since it relies on the fact that bi-implication is both commutative and associative [36]:

$$((\exists x.\forall y.P(x) \Leftrightarrow P(y)) \Leftrightarrow ((\exists x.Q(x)) \Leftrightarrow (\forall y.Q(y)))) \Leftrightarrow$$
$$((\exists x.\forall y.Q(x) \Leftrightarrow Q(y)) \Leftrightarrow ((\exists x.P(x)) \Leftrightarrow (\forall y.P(y))))$$

Comparing the declarative proofs in Isabelle/HOL and SPA is a good exercise for the students.

In addition to our tools for teaching logic we cover the following online papers:

1. M. Ben-Ari (2020): A Short Introduction to Set Theory [4]
2. W. M. Farmer (2008): The Seven Virtues of Simple Type Theory [6]
3. T. C. Hales (2008): Formal Proof [17]
4. T. Nipkow (2021): Programming and Proving in Isabelle/HOL [25]
5. L. C. Paulson (2018): Computational Logic: Its Origins and Applications[31]

The paper by Farmer provides a concise definition of higher-order logic and the tutorial by Nipkow provides a substantial set of exercises which the students must solve.

6   Discussion and Future Work

As mentioned, our BSc course uses our Sequent Calculus Verifier (SeCaV), which is embedded in Isabelle/HOL, for several exercise sessions and assignments. While the system is designed to be quite simple to use and understand, we have experienced that some students have a hard time writing proofs in the system. Additionally, the embedding in Isabelle/HOL is not able to give very helpful error messages if a proof is wrong. To alleviate these issues, we have recently developed an online tool called the SeCaV Unshortener [11], which allows users to write proofs in a simpler syntax, which is then automatically translated into the embedding in Isabelle. Additionally, the tool is able to warn users about mistakes in their proofs by explicitly telling users why e.g. a proof rule cannot be applied. Recent research indicates that this kind of feedback impacts learning in computer science significantly, and is sufficient to allow students to move forward in most cases [18].

We also use SeCaV in our MSc course but only as self-study concerning the course prerequisites and selected parts of the papers [11,12,15] in the first weeks of the course.

We would like to integrate even more algorithms and proofs into Isabelle. Work is currently ongoing on an Isabelle implementation and proof of correctness of a tool for converting formulas to conjunctive normal form.

Michaelis and Nipkow [23] formalized a number of proof systems for propositional logic in Isabelle/HOL: resolution, natural deduction, sequent calculus and an axiomatic system. We would like to extend this line of work to first-order logic and higher-order logic.

We find that one of the main issues in both our 2020 and 2021 course on automated reasoning and formally verified functional programming is the course prerequisites. Functional programming is a prerequisite but we do not require a specific language and it is not possible to exclude any students. This is a real problem and in general we need to use the first part of the course to teach some of the prerequisites. Another prerequisite is mathematical logic—syntax, semantics and proof systems—and we use the micro provers to teach logic, functional programming and the basics of a proof assistant, in particular Isabelle, in a way that is challenging to almost all students. It is not for beginners and some students will most likely quit the course in the first month. In 2021, after the first month, 37 students were active and almost everyone submitted the first assignment. We have no clear solution to the issues concerning the course prerequisites but for 2022 we plan to offer a series of online sessions for self-study in mathematical logic and functional programming.

7   Conclusion

We have presented our detailed experiences teaching two courses on formal methods. The first course is the bachelor course on logical systems and logic programming, which has a focus on classical first-order logic and automatic theorem proving. We have additionally described how we use Prolog and Isabelle to introduce students to logic and formal methods.

The second course is the master course on automated reasoning, which has a focus on classical higher-order logic and interactive theorem proving. The proof assistant Isabelle is used more heavily in this course, and we use Isabelle/Pure as well as Isabelle/HOL. We have also described our online tools to be used with Isabelle/HOL, in particular the Sequent Calculus Verifier (SeCaV) and the Natural Deduction Assistant (NaDeA). In addition, we have described our innovative Students' Proof Assistant which is formally verified in Isabelle/HOL and integrated in Isabelle/jEdit using Isabelle/ML.

We have described how our courses fit into the overall computer science and engineering curriculum, and what issues and challenges we experience that students often face. We have suggested some future work on the courses by which we hope to improve student learning outcomes.

Our teaching philosophy is related to the IsaFoL (Isabelle Formalization of Logic) project [5] which aims at developing formalizations in Isabelle/HOL of logics, proof systems, and automatic/interactive provers. Notable work in the same line includes the soundness and completeness of epistemic [8] and hybrid [9] logic and an ordered resolution prover for first-order logic [35]. These formalizations can serve as starting point for a student project to formalize the soundness and completeness of various other proof systems and provers.

We would like to formalize even more topics within basic logic such that students can explore concrete and executable definitions of various topics such as Skolemization while also seeing formal proofs of their correctness. Our overall conclusion is that using formal methods, in particular the proof assistant Isabelle, as a central tool for teaching logic and formal methods is possible as we have demonstrated since our first use of the Natural Deduction Assistant (NaDeA) and the Sequent Calculus Verifier (SeCaV) in 2014 and 2019, respectively.

Acknowledgements. We thank Asta Halkjær From for comments on drafts. We thank the three anonymous reviewers whose comments and suggestions helped improve the paper.

References

1. Baanen, A., Bentkamp, A., Blanchette, J., Limperg, J., Hölzl, J.: The Hitchhiker's Guide to Logical Verification (2020). https://github.com/blanchette/logical_verification_2020
2. Bella, G.: You already used formal methods but did not know it. In: Dongol, B., Petre, L., Smith, G. (eds.) FMTea 2019. LNCS, vol. 11758, pp. 228–243. Springer, Cham (2019). https://doi.org/10.1007/978-3-030-32441-4_15
3. Ben-Ari, M.: Mathematical Logic for Computer Science. Springer, London (2012)
4. Ben-Ari, M.: A Short Introduction to Set Theory (2020). https://www.weizmann.ac.il/sci-tea/benari/sites/sci-tea.benari/files/uploads/books/set.pdf
5. Blanchette, J.C.: Formalizing the metatheory of logical calculi and automatic provers in Isabelle/HOL (invited talk). In: Mahboubi, A., Myreen, M.O. (eds.) Proceedings of the 8th ACM SIGPLAN International Conference on Certified Programs and Proofs, CPP 2019, Cascais, Portugal, 14–15 January 2019, pp. 1–13. ACM (2019)
6. Farmer, W.M.: The seven virtues of simple type theory. J. Appl. Log. **6**(3), 267–286 (2008). https://doi.org/10.1016/j.jal.2007.11.001
7. From, A.H.: Formalizing Henkin-style completeness of an axiomatic system for propositional logic. In: Proceedings of the Web Summer School in Logic, Language and Information (WeSSLLI) and the European Summer School in Logic, Language

and Information (ESSLLI) Virtual Student Session, pp. 1–12 (2020). Preliminary paper, accepted for Springer post-proceedings

8. From, A.H.: Epistemic logic: completeness of modal logics. Archive of Formal Proofs, October 2018. https://devel.isa-afp.org/entries/Epistemic_Logic.html, Formal proof development

9. From, A.H.: Formalizing a Seligman-style tableau system for hybrid logic. Archive of Formal Proofs, December 2019. https://isa-afp.org/entries/Hybrid_Logic.html, Formal proof development

10. From, A.H., Eschen, A.M., Villadsen, J.: Formalizing axiomatic systems for propositional logic in Isabelle/HOL. In: Kamareddine, F., Sacerdoti Coen, C. (eds.) CICM 2021. LNCS (LNAI), vol. 12833, pp. 32–46. Springer, Cham (2021). https://doi.org/10.1007/978-3-030-81097-9_3

11. From, A.H., Jacobsen, F.K., Villadsen, J.: SeCaV: a sequent calculus verifier in Isabelle/HOL. In: 16th International Workshop on Logical and Semantic Frameworks with Applications (LSFA 2021) – Presentation Only/Online Papers, pp. 1–16 (2021). https://mat.unb.br/lsfa2021/pages/lsfa2021_proceedings/LSFA_2021_paper_5.pdf

12. From, A.H., Jensen, A.B., Schlichtkrull, A., Villadsen, J.: Teaching a formalized logical calculus. Electron. Proc. Theor. Comput. Sci. **313**, 73–92 (2020). https://doi.org/10.4204/EPTCS.313.5

13. From, A.H., Lund, S.T., Villadsen, J.: A case study in computer-assisted meta-reasoning. In: González, S.R., Machado, J.M., González-Briones, A., Wikarek, J., Loukanova, R., Katranas, G., Casado-Vara, R. (eds.) DCAI 2021. LNNS, vol. 332, pp. 53–63. Springer, Cham (2022). https://doi.org/10.1007/978-3-030-86887-1_5

14. From, A.H., Villadsen, J.: Teaching automated reasoning and formally verified functional programming in Agda and Isabelle/HOL. In: 10th International Workshop on Trends in Functional Programming in Education (TFPIE 2021) – Presentation Only/Online Papers, pp. 1–20 (2021). https://wiki.tfpie.science.ru.nl/TFPIE2021

15. From, A.H., Villadsen, J., Blackburn, P.: Isabelle/HOL as a meta-language for teaching logic. Electron. Proc. Theor. Comput. Sci. **328**, 18–34 (2020). https://doi.org/10.4204/eptcs.328.2

16. Grover, S.: Toward a framework for formative assessment of conceptual learning in K-12 computer science classrooms. In: Proceedings of the 52nd ACM Technical Symposium on Computer Science Education, SIGCSE 2021, pp. 31–37 (2021). https://doi.org/10.1145/3408877.3432460

17. Hales, T.C.: Formal proof. Not. Am. Math. Soc. **55**, 1370–1380 (2008)

18. Hao, Q., et al.: Towards understanding the effective design of automated formative feedback for programming assignments. Comput. Sci. Educ. 1–23 (2021). https://doi.org/10.1080/08993408.2020.1860408

19. Harrison, J.: Formalizing basic first order model theory. In: Grundy, J., Newey, M. (eds.) TPHOLs 1998. LNCS, vol. 1479, pp. 153–170. Springer, Heidelberg (1998). https://doi.org/10.1007/BFb0055135

20. Harrison, J.: Handbook of Practical Logic and Automated Reasoning. Cambridge University Press, Cambridge (2009)

21. Jensen, A.B., Larsen, J.B., Schlichtkrull, A., Villadsen, J.: Programming and verifying a declarative first-order prover in Isabelle/HOL. AI Commun. **31**(3), 281–299 (2018)

22. Krauss, A.: Defining Recursive Functions in Isabelle/HOL (2021). https://isabelle.in.tum.de/doc/functions.pdf

23. Michaelis, J., Nipkow, T.: Formalized proof systems for propositional logic. In: Abel, A., Forsberg, F.N., Kaposi, A. (eds.) 23rd International Conference on Types for Proofs and Programs, TYPES 2017, Budapest, Hungary, 29 May–1 June 2017. LIPIcs, vol. 104, pp. 5:1–5:16. Schloss Dagstuhl - Leibniz-Zentrum für Informatik (2017)

24. Nipkow, T.: Teaching semantics with a proof assistant: no more LSD trip proofs. In: Kuncak, V., Rybalchenko, A. (eds.) VMCAI 2012. LNCS, vol. 7148, pp. 24–38. Springer, Heidelberg (2012). https://doi.org/10.1007/978-3-642-27940-9_3

25. Nipkow, T.: Programming and Proving in Isabelle/HOL (Tutorial) (2021). https://isabelle.in.tum.de/doc/prog-prove.pdf

26. Nipkow, T.: Teaching algorithms and data structures with a proof assistant (invited talk). In: Hritcu, C., Popescu, A. (eds.) 10th ACM SIGPLAN International Conference on Certified Programs and Proofs, Virtual Event, CPP 2021, Denmark, 17–19 January 2021, pp. 1–3. ACM (2021). https://doi.org/10.1145/3437992.3439910

27. Nipkow, T., et al.: Functional Algorithms, Verified! (2021). https://functional-algorithms-verified.org/

28. Nipkow, T., Eberl, M., Haslbeck, M.P.L.: Verified textbook algorithms. In: Hung, D.V., Sokolsky, O. (eds.) ATVA 2020. LNCS, vol. 12302, pp. 25–53. Springer, Cham (2020). https://doi.org/10.1007/978-3-030-59152-6_2

29. Nipkow, T., Klein, G.: Concrete Semantics - With Isabelle/HOL. Springer, Heidelberg (2014)

30. Nipkow, T., Wenzel, M., Paulson, L.C. (eds.): Isabelle/HOL. LNCS, vol. 2283. Springer, Heidelberg (2002). https://doi.org/10.1007/3-540-45949-9

31. Paulson, L.C.: Computational logic: its origins and applications. Proc. R. Soc. A. **474**(2210), 20170872 (2018). https://doi.org/10.1098/rspa.2017.0872

32. Peltier, N.: A variant of the superposition calculus. Archive of Formal Proofs, September 2016. http://isa-afp.org/entries/SuperCalc.shtml, Formal proof development

33. Pierce, B.C., et al.: Software Foundations – 6 Online Textbooks (2021). https://softwarefoundations.cis.upenn.edu/

34. Reis, G.: Facilitating meta-theory reasoning (invited paper). In: Pimentel, E., Tassi, E. (eds.) Proceedings Sixteenth Workshop on Logical Frameworks and Meta-Languages: Theory and Practice, Pittsburgh, USA, 16th July 2021. Electronic Proceedings in Theoretical Computer Science, vol. 337, pp. 1–12. Open Publishing Association (2021). https://doi.org/10.4204/EPTCS.337.1

35. Schlichtkrull, A., Blanchette, J., Traytel, D., Waldmann, U.: Formalizing Bachmair and Ganzinger's ordered resolution prover. J. Autom. Reason. **64**(7), 1169–1195 (2020)

36. Schlichtkrull, A., Villadsen, J., From, A.H.: Students' Proof Assistant (SPA). In: Quaresma, P., Neuper, W. (eds.) Proceedings 7th International Workshop on Theorem Proving Components for Educational Software, ThEdu@FLoC 2018, Oxford, United Kingdom, 18 July 2018. Electronic Proceedings in Theoretical Computer Science, vol. 290, pp. 1–13. Open Publishing Association (2018). https://doi.org/10.4204/EPTCS.290.1

37. Villadsen, J.: A micro prover for teaching automated reasoning. In: Seventh Workshop on Practical Aspects of Automated Reasoning (PAAR 2020) – Presentation Only/Online Papers, pp. 1–12 (2020). https://www.eprover.org/EVENTS/PAAR-2020.html

38. Villadsen, J.: Tautology checkers in Isabelle and Haskell. In: Calimeri, F., Perri, S., Zumpano, E. (eds.) Proceedings of the 35th Edition of the Italian Conference on Computational Logic (CILC 2020), Rende, Italy, 13–15 October 2020. CEUR Workshop Proceedings, vol. 2710, pp. 327–341. CEUR-WS.org (2020). http://ceur-ws.org/Vol-2710/paper-21.pdf
39. Villadsen, J., From, A.H., Schlichtkrull, A.: Natural Deduction Assistant (NaDeA). In: Quaresma, P., Neuper, W. (eds.) Proceedings 7th International Workshop on Theorem Proving Components for Educational Software, THedu@FLoC 2018, Oxford, United Kingdom, 18 July 2018. EPTCS, vol. 290, pp. 14–29 (2018). https://doi.org/10.4204/EPTCS.290.2
40. Villadsen, J., Schlichtkrull, A., From, A.H.: A verified simple prover for first-order logic. In: Konev, B., Urban, J., Rümmer, P. (eds.) Proceedings of the 6th Workshop on Practical Aspects of Automated Reasoning (PAAR 2018) co-located with Federated Logic Conference 2018 (FLoC 2018), Oxford, UK, 19 July 2018. CEUR Workshop Proceedings, vol. 2162, pp. 88–104. CEUR-WS.org (2018). http://ceur-ws.org/Vol-2162/paper-08.pdf
41. Wenzel, M.: The Isabelle/Isar Reference Manual (2021). https://isabelle.in.tum.de/doc/isar-ref.pdf
42. Wenzel, M.: Isar—a generic interpretative approach to readable formal proof documents. In: Bertot, Y., Dowek, G., Théry, L., Hirschowitz, A., Paulin, C. (eds.) TPHOLs 1999. LNCS, vol. 1690, pp. 167–183. Springer, Heidelberg (1999). https://doi.org/10.1007/3-540-48256-3_12

Introducing Formal Methods to Students Who Hate Maths and Struggle with Programming

Nisansala Yatapanage[✉]

School of Computing, Australian National University, Canberra, Australia
yatapanage@acm.org

Abstract. Formal methods is usually considered a subject that requires a strong ability in mathematics. For this reason, the subject is generally taught to higher level undergraduate or postgraduate students. However, introducing formal methods at early undergraduate levels provides students with a solid foundation and increases their natural inclination to use formal methods in their future careers. This paper describes an attempt to introduce formal methods concepts to early undergraduate students who have weak mathematical knowledge and are only beginning to learn programming concepts. The aim was to provide the students with a gentle introduction without focussing on the complex mathematical aspects. Instead, the material was presented as the core ideas combined with practical and problem-solving tasks. The results were that the students generally found the concepts easy to understand, despite them struggling with the theory behind other Computer Science topics.

1 Introduction

It is clearly necessary to teach formal methods in a Computer Science undergraduate curriculum, in order to ensure that the students graduate with the skills needed to apply formal methods to their future work. However, it is usually assumed that a good background in mathematics is required for students to understand the concepts of formal methods. Formal methods courses are usually placed in advanced undergraduate or postgraduate levels. Instead, introducing the concepts at much earlier levels integrates the topic more strongly into the core foundations of the students' computer science education. Teaching formal methods together with introductory logic and programming courses allows students to naturally integrate formal methods ideas into all of their program designs, while they are still learning programming.

This paper explores some of the ideas and insights gained from the author's experiences teaching undergraduate courses containing formal methods, with particular focus on a 2nd-year concurrency course taught at De Montfort University, U.K.

The ideas discussed in this paper are from a course that the author taught at her previous institution, De Montfort University, Leicester, U.K.

© Springer Nature Switzerland AG 2021
J. F. Ferreira et al. (Eds.): FMTea 2021, LNCS 13122, pp. 133–145, 2021.
https://doi.org/10.1007/978-3-030-91550-6_10

Some of the students of this course were unfamiliar with even higher level high school mathematics concepts, such as logarithms, and were struggling with general programming concepts such as recursion. The author realised this during the first week of classes, finding that while there were some students who were comfortable with Java programming, many others were confused by simple concepts. An early data structures lecture on asymptotic analysis revealed the students' level of mathematics knowledge. This situation is not unique to De Montfort University. As pointed out by [MOPD19], there is a trend towards removing mathematics as a prerequisite for Computer Science bachelor degrees. While these 2nd-year students had studied basic programming and some mathematics in their first year, it was clear that many of them still required further training in these areas. The question was, could (and should) these students be taught formal methods topics? Would it be better to wait until the students had studied more programming and mathematics, and introduce formal methods in their third year?

It turned out that formal methods *can* be taught at this level. It just depends on how the subject matter is taught. The basic ideas of formal methods can be conveyed without requiring any knowledge of complex mathematics. It is further suggested that formal methods could be taught even before university level, giving younger students a taste of the general ideas while they learn simple programming concepts in high school. The author gave a lecture demonstrating model checking to Year 10 (in the U.K. system) high school students. While this attempt was unsuccessful, lessons learnt from this are also discussed.

2 Introducing Formal Methods into a Concurrency Course

An important question is where formal methods should be taught in an early undergraduate curriculum. The concurrency course at De Montfort University developed by the author covered the basics of concurrency concepts, such as mutual exclusion, deadlock, semaphores, etc. This was an ideal course in which to introduce formal methods to the students. Concurrency fits naturally with formal methods, as a significant focus of formal methods research is on handling concurrent programs.

Why are We Learning This?

A student once asked the lecturer this question. While most students won't bother to ask, most are thinking it. Clear motivations for *why* it is important to learn formal methods are essential. Concurrency provides a useful strategy for this. Small concurrent programs are notoriously difficult to write correctly. It is easy to demonstrate to the students the benefits of using formal methods to find errors in a concurrent program that they had assumed to be correct. For example, a "simple" concurrent program such as the one in Fig. 1 is complex enough to challenge a student who has just begun learning concurrency. Note that instead of using $n = n + 1$, splitting it into two statements, where the value of n is first stored into a local variable, helps the students to understand that the value of n could have changed by the time the process attempts to increment it. After showing the students that these simple programs can lead to a variety of unexpected results, when the concepts of formal methods are later introduced, the students can see the value in methods that help them to reason about such programs.

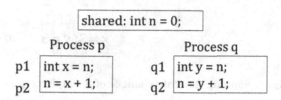

Fig. 1. A *Simple* concurrent program

The course was structured to begin with teaching the students the essential concurrency concepts, and then introduced students to formal methods ideas, in particular temporal logic, model checking and rely/guarantee reasoning.

Explain the Concepts at the Level of the Students

It must be kept in mind that the students were fairly new to programming. They had learnt the basics in their first year, but were learning Java for the first time in a course running in parallel. Therefore, the concepts of concurrency and formal methods had to be taught in a simple way, without focussing on the underlying mathematics which would have scared many students away.

Linear Temporal Logic (LTL) [Pnu77] was chosen as it is easy to understand its semantics, compared to some other logics such as Computation Tree Logic (CTL) [CE81], for which the branching structure can be harder for a beginner to understand.

Before showing the students LTL, they had to first understand some concepts that might be rushed through in a more advanced level course. The students at this level had never encountered the idea of *states* of a system. To understand how LTL properties hold over states, they had to first understand what states are. This was explained to the students using a simple program and stepping through the value of the state after each step. Figure 2 demonstrates this. Note that the branch corresponding to the *else* case of the *if* statement has not been shown on this diagram, but was explained later in the course.

There are a few key points that must be mentioned to the students. One is that the states are *between* the program statements, as the statements are the transitions between states. This is a new concept that many students would not realise. It is also a concept that could lead to a lot of confusion later on,

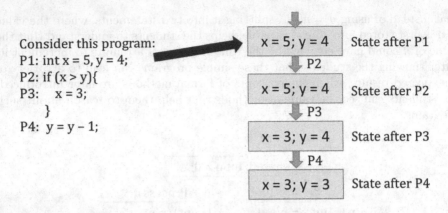

Consider this program:
P1: int x = 5, y = 4;
P2: if (x > y){
P3:   x = 3;
   }
P4: y = y − 1;

x = 5; y = 4   State after P1
P2
x = 5; y = 4   State after P2
P3
x = 3; y = 4   State after P3
P4
x = 3; y = 3   State after P4

Fig. 2. Showing the state changes of a simple program

when trying to understand logics holding over the states. Another point is that some states in the diagram have the same values for variables, but are actually different states, because they also implicitly contain program counters indicating where the program is up to. These are both points that advanced courses would not focus on. When explaining new concepts to students at beginner levels, it is necessary to think carefully about what key points would be confusing to the students if overlooked.

Next, slightly more complex programs, particularly ones which lead to branching in the corresponding transition system, were shown. The different LTL operators were then each shown on a timeline to explain their behaviour and gradually more complex formulas were shown. For example, Fig. 3 shows a timeline demonstrating how the **F** (future) operator works. By going through each operator separately in lectures, students were able to easily grasp their behaviour. As the lectures progressed, the operators were combined into various common styles of formulas, such as $\mathbf{G}(p \implies \mathbf{F}(q))$.

Compare this with a model checking lecture the author gave at another university as part of an undergraduate Software Engineering course, where the students all had a strong mathematics background. For that group of students, the entire set of LTL operators, as well as the basics of model checking, could be covered in a single lecture. It is important to teach at the pace required by the students in the class.

The lectures were supplemented with tutorials, where the students were given a set of questions on LTL. These questions began with simple definitions of the operators, presented as a match-the-answer style of question, where LTL operators had to be matched with their definitions in words. This was followed by various properties which the students had to evaluate using a given state transition diagram, such as in Fig. 4.

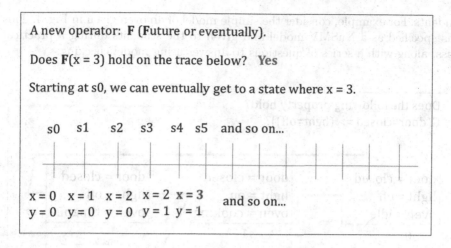

A new operator: **F** (Future or eventually).

Does **F**(x = 3) hold on the trace below?   **Yes**

Starting at s0, we can eventually get to a state where x = 3.

Fig. 3. Lecture slide showing a timeline demonstrating how the F operator works

Leave out Complex Mathematical Descriptions

Students who are not strong in mathematics will quickly give up if presented with formulae that look complicated. Often, they will not wait around to hear the explanation given by the lecturer which would have made it clear that the formulae are not as scary as they first appear. Therefore, it is best to simply leave out the underlying mathematics at this level, until the students have grasped the concepts on a practical level. Then, later, more advanced courses can cover the mathematics aspects and students will have a better chance at understanding them, as they have already been familiar with using the concepts practically. For example, the formal semantics of LTL are unnecessary for the students at this time. Practical tasks and examples allow them to learn how the concepts work, without having to be slowed down by the underlying theory.

What Made You Interested in Formal Methods?

It is a question many researchers may have forgotten the answer to: what attracted them to formal methods in the first place. Perhaps it was working on a particular verification task that required problem-solving. While every student is different, many would find working on a realistic problem more interesting than reading pages of mathematical proofs. However, the art is in finding a realistic problem that students at early levels can handle.

Model checkers [QS82,CE81] are useful teaching tools[1], as it is generally simple to explain to the students how to run them, and they give a clear answer, at least on small problems that do not run into state explosion. Students can be given a small problem to model check. The difficult aspect is usually creating the model, so providing an existing model makes the process much easier for the

---

[1] For a useful textbook on model checking, see [BK08].

students. For example, consider the simple model of an oven given in Fig. 4. This was specified as a NuSMV model and given to the students during a practical class, along with a series of questions to answer using model checking.

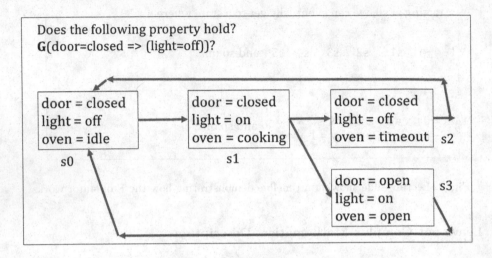

Fig. 4. A simple case study for model checking

The students were given detailed steps on how to use the model checker. They were not asked straight away to start verifying LTL properties. Instead, the tasks started with running NuSMV's interactive mode and stepping through the states. This helped them to see how the states changed with each transition, complementing the lecture about states and traces.

Next, the task was to try verifying certain given properties, e.g. $\mathbf{G}(door = closed \implies \mathbf{F}(oven = cooking)$. The students were encouraged to try to think of the answer before running the model checker. It had already been explained in the lectures which types of properties would return counterexamples with cycles.

The following task was to make certain changes to the model and then re-run the same properties. They were also asked to create properties of their own, e.g.

– *Write a property that holds on the original model but not on the new model.*
– *Write a property containing two nested X operators.*

The fun part of model checking, which would interest and motivate students, is to debug problems in a model. A mistake was deliberately introduced into the model and the students were asked to attempt to locate the issue by studying the results given by the model checker for various properties. It was difficult to devise a mistake that was simple enough for the students to find and yet was not obvious when just looking at the model specification. Unfortunately, the chosen error ended up being easy to spot, but enthusiastic students still seemed interested in using the model checker to find it.

The course had weekly practicals, made up of exercises related to the lecture topics. Students were encouraged to work through the exercises at their own pace. The practicals at the start of the semester focussed on concurrent programming topics, while later practicals were on model checking. While the students who completed these tasks seemed to enjoy model checking and improved their understanding of LTL, students had to first complete the exercises on concurrent programming. As many of the students were finding programming difficult, and also due to an unfortunate low attendance in the lab classes, only some of the students attempted the model checking tasks.[2] Rearranging the order of the practical tasks may have helped, but that would have meant moving the entire formal methods section to earlier in the concurrency course. Another solution could be to include a model checking task in the assignment, which was not done during this run of the course.

Introduce a Variety of Formal Methods Approaches

While model checking and LTL were good topics for introducing the students to verification, it is important to show the students that there are many different approaches. For this course, rely/guarantee reasoning [Jon83a, Jon83b] was chosen as another topic, in order to introduce students to program reasoning approaches for concurrency. The basic idea of rely/guarantee reasoning is very straight-forward and easy for a beginner to grasp.

First, the concepts of *pre conditions* and *post conditions* were explained, along with several examples. Most of one entire lecture was on just the pre and post conditions. Then, the ideas of rely and guarantee conditions were introduced. The meaning of rely and guarantee conditions were explained using a diagram such as in Fig. 5.

Then, simple concurrent programs were shown, along with rely and guarantee conditions for each process. The concept that the guarantee conditions of each process should satisfy the rely condition of the other was a straight-forward, clear concept and most students found questions such as the one in Fig. 6 easy to understand.

Realistic Examples

As discussed previously, the students wanted to see the relevance of what they were studying. A good way to show them the relevance of formal methods is to discuss a realistic case study. The students were shown a case study of an industrial metal press, as shown in [GLYW05]. A simple animation was used to show the students how the metal press operated. A set of safety properties were then given in LTL, which was a nice way for the students to see how LTL properties are useful in practice. For example, the following properties, taken from [GLYW05] were discussed:

---

[2] At the time of teaching the course, low attendance was a common problem across the School. Students often chose not to attend any classes, regardless of the course or topic.

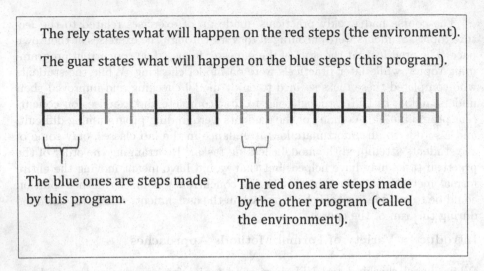

The rely states what will happen on the red steps (the environment).

The guar states what will happen on the blue steps (this program).

The blue ones are steps made by this program.

The red ones are steps made by the other program (called the environment).

Fig. 5. Lecture slide with an overview of Rely and Guarantee Conditions. (Note that *guarantee* has been shortened to *guar* to match with the convention used in specifications.)

Recap: A **rely** tells us what the **other** program should do.
A **guar** tells us what **this** program guarantees to do.

Can the two programs below run concurrently together?

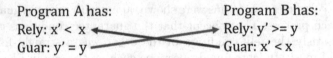

Program A has:
Rely: x' < x
Guar: y' = y

Program B has:
Rely: y' >= y
Guar: x' < x

Yes, because A's guar satisfies B's rely and B's guar satisfies A's rely.

Fig. 6. A rely/guarantee example.

– $\mathbf{G}((plunger = fallingFast) \implies (motor = off))$
– $\mathbf{G}((plunger = atTop \wedge operator = releasedButton) \implies (motor = on))$

The results of the model checking analysis were discussed, including explaining the relevant counterexamples. It was interesting to note that even though this lecture was given at the end of the semester and the students knew that this was not on the final exam, there was still relatively good attendance, showing

that students are keen to see realistic examples. Tavolato and Vogt [TV12] stress
the importance of showing students the practical relevance in formal methods,
using real-life case studies.

3 Assessments - Traditional and Online

The LTL and rely/guarantee topics taught in the concurrency course were
assessed as part of an online Blackboard exam. In a previous year, similar questions had been assessed as a written paper exam in another related course taught
by the author (which was later adapted into the concurrency course). The written exam questions had to therefore be modified into an online format.

In the paper exam, the questions on LTL were presented in a similar way to
Fig. 4. Students were given a transition system such as in the figure, and given
a set of LTL formulas. For each one, they were asked:

- Does the property hold? (Yes/No)
- Describe the property in words.
- If the property does not hold, give a counterexample as a trace showing the
  states reached after each step. Then explain the counterexample in words.
- If the property holds, explain why in words, referring to the states.

The idea of asking the students to explain the properties in words was to
give them the opportunity to demonstrate some understanding even if they were
unable to write correct counterexamples. Additionally, this also helped to determine whether a student had simply guessed an answer without actually understanding it.

This type of question was adapted into the online format by turning it into
a multiple-choice question. The students had to select from a given set of counterexamples, which included the option that the property is true. A separate
multiple-choice question asked the meaning of the property, giving a set of possible answers to choose from. The obvious difficulty is that students could just
guess the answers, unlike the paper version of the exam. To help in resolving this,
the answers were chosen so that a student who did not understand the property
would not be able to easily guess the correct answer. They could still randomly
select the correct one, but this could not be avoided in a multiple-choice exam.
The rely/guarantee questions were posed in a similar way to Fig. 6, with multiple
choices, asking the students whether the two processes satisfy each other's rely
conditions.

Generally, the students seemed to find the multiple-choice version of the
exam much easier. It seemed that the students found having the choices easier
than coming up with the solutions from scratch. However, it is difficult to make
conclusions because the previous course structure was significantly different and
therefore, it may have been those changes that helped the students to understand
the topic better.

The other parts of the exam were on the other concurrency topics, but were
also related to the formal methods aspects, as students were asked to write out

traces. For example, a question gave the students a small concurrent program such as in Fig. 1 and asked them to write out two traces that result in different values for the shared variable n. The students were generally able to answer this question correctly and it was clear that understanding what traces meant helped them to understand the LTL properties.

Overall, it was interesting to find that the students found the formal methods topics easier than some of the other Computer Science topics they were learning, such as data structures and algorithms.

4   Formal Methods Before University

Computer Science is now a part of the high school level curriculum in many places, at least as an option for students to choose. There is therefore no reason why formal methods couldn't be introduced at high school level as well. Integrating early Computer Science education with formal methods would give students a solid foundation. The author gave a lecture to Year 10 students as a way of introducing students to university education. The lecture gave a basic overview of the ideas of verification and why it is useful. It included a demonstration of a model checker finding an error in a system. The same Industrial Press case study was used as the example, but with its operation simplified and only discussed as an overview.

Unfortunately, the students were generally not very interested. However, it was a general group of students which included students who had not been studying Computer Science. The example demonstrated was probably too complex for those who had not even experienced basic programming. It is suggested that a carefully planned introduction to formal methods could be accomplished as part of high school study. The ideas used for teaching formal methods to the university students who struggled with mathematics would be applicable in this setting as well. A single lecture appeared to be too fast for the high school students, and tended to lose their concentration. A series of practical lessons where the students are given simple exercises in logic or fun simplified problems to try in a tool would be more effective. They would then have been exposed to the general ideas long before studying formal methods in depth at university.

Teaching formal methods at the school level has been proposed by [MOPD19], where they used interesting problems to teach school children the formal methods style of thinking. They found that the same principles applied for teaching first-year students formal methods.

5   Related Work

[Mor19] advocates similar ideas as suggested in this paper, of introducing formal methods at early levels along with programming. [PW18] discuss the benefits of teaching formal verification together with concurrency. When first learning concurrency, students tend to find it difficult to understand the challenges involved. [LV05] discuss the fact that students tend to find it easy to recall definitions on concurrency, but struggle with creating concurrent programs without common errors. Showing students techniques for verifying the correctness of concurrent algorithms gives them a good appreciation for the complexity produced by concurrency. A study by [FB97] concludes that concurrency can be taught to lower undergraduate students, despite often appearing to be more suitable as an advanced topic. [Kra07] discusses using model checking to teach students concurrency and the need to teach students how to develop suitable abstractions.

[Win00] encourages teaching students verification at early undergraduate levels to give them a solid basis in the area, allowing them to feel naturally inclined to incorporate such techniques when they enter the workforce, and advocates using tool support to teach students formal methods. Recent formal methods research has been increasingly focussing on providing tool support that is easy to use by the average engineer, rather than experts in formal methods. These approaches are simple and easy for students to quickly grasp. [BS08] discuss the benefits they observed when using a practical verification tool for a formal methods course, finding it to be a useful way to introduce students to formal methods. While they discuss some potential drawbacks, such as students sometimes relying on the tool without fully understanding the underlying logic, overall they find it to be beneficial. [Zin08] describes another similar approach using lightweight formal methods, allowing students to learn the concepts without having to understand complex mathematical proofs. [DR19] found that using the *Krakatoa* tool for teaching students improved their understanding. [Liu02] points out that incorporating model checking into an undergraduate course provides a link between education and industry, as model checking is a useful skill in industry.

[MOPD19] found that teaching university students formal methods in their first year required a set of principles which are similar to those suggested in this paper. They propose using interesting and fun riddles and games, leaving formal definitions until after the students have understood the concepts informally and using interactive problem-solving groups. They show that even complex concepts such as bisimulation can be taught by relating them to interesting problems and games, without showing students the complex mathematical theory.

6 Conclusion

There are clear advantages in teaching students formal methods at early stages of undergraduate study. The concurrency course discussed in this paper has demonstrated that it is indeed possible to introduce students to formal methods even when they have weak mathematical backgrounds and are still learning programming. Integrating formal methods with early Computer Science courses helps to ensure that the students understand the benefits of formal methods. By focussing on problem-solving tasks instead of the underlying mathematical theory, students find the topics easy to understand and are less likely to be scared off. It is important to teach at the appropriate pace for the students and to carefully explain all aspects that they might struggle with. Similar approaches may be possible for teaching students at pre-university levels. Introducing formal methods into high school study while learning basic programming would help students to have a strong foundation.

Acknowledgements. The author would like to thank David Smallwood and Luke Attwood for their support and advice during the design of the concurrency course.

References

[BK08]   Baier, C., Katoen, J.-P.: Principles of Model Checking. MIT Press, Cambridge (2008)

[BS08]   Boyatt, R.C., Sinclair, J.E.: Experiences of teaching a lightweight formal method. In: Formal Methods in Computer Science Education 2008 Workshop, Proceedings, pp. 71–80 (2008)

[CE81]   Clarke, E.M., Emerson, E.A.: Design and synthesis of synchronization skeletons using branching time temporal logic. In: Kozen, D. (ed.) Logic of Programs 1981. LNCS, vol. 131, pp. 52–71. Springer, Heidelberg (1982). https://doi.org/10.1007/BFb0025774

[DR19]   Divasón, J., Romero, A.: Using krakatoa for teaching formal verification of java programs. In: Dongol, B., Petre, L., Smith, G. (eds.) FMTea 2019. LNCS, vol. 11758, pp. 37–51. Springer, Cham (2019). https://doi.org/10.1007/978-3-030-32441-4_3

[FB97]   Feldman, M.B., Bachus, B.D.: Concurrent programming CAN be introduced into the lower-level undergraduate curriculum. In: Cassel, L.N., Daniels, M., Miller, J.E., Davies, G. (eds.) 2nd Annual Conference on Integrating Technology into Computer Science Education, ITiCSE 1997, Proceedings, pp. 77–79. ACM (1997)

[GLYW05] Grunske, L., Lindsay, P., Yatapanage, N., Winter, K.: An automated failure mode and effect analysis based on high-level design specification with behavior trees. In: Romijn, J., Smith, G., van de Pol, J. (eds.) IFM 2005. LNCS, vol. 3771, pp. 129–149. Springer, Heidelberg (2005). https://doi.org/10.1007/11589976_9

[Jon83a] Jones, C.B.: Specification and design of (parallel) programs. In: Proceedings of IFIP 1983, North-Holland, pp. 321–332 (1983)

[Jon83b] Jones, C.B.: Tentative steps toward a development method for interfering programs. ACM ToPLaS **5**(4), 596–619 (1983)

[Kra07]   Kramer, J.: Is abstraction the key to computing? Commun. ACM **50**(4), 36–42 (2007)

[Liu02]   Liu, H.: A proposal for introducing model checking into an undergraduate software engineering curriculum. In: CCSC Southeastern Conference December 2002, Proceedings, pp. 259–270 (2002)

[LV05]   Lutz, M., Vallino, J.: Concurrent system design: applied mathematics & modeling in software engineering education (2005)

[MOPD19] Moller, F., O'Reilly, L., Powell, S., Denner, C.: Teaching them early: formal methods in school. In: Proceedings of FMFun'2019 Formal Methods - Fun for Everybody, (2019, to appear)

[Mor19]   Morgan, C.: Is formal methods really essential? (invited talk). In: Formal Methods Teaching Workshop and Tutorial, FMTea19 (2019)

[Pnu77]   Pnueli, A.: The temporal logic of programs. In: 18th Annual Symposium on Foundations of Computer Science, pp. 46–57. IEEE (1977)

[PW18] Pedersen, J.B., Welch, P.H.: The symbiosis of concurrency and verification: teaching and case studies. Formal Aspects Comput. **30**(2), 239–277 (2017). https://doi.org/10.1007/s00165-017-0447-x

[QS82] Queille, J.P., Sifakis, J.: Specification and verification of concurrent systems in CESAR. In: Dezani-Ciancaglini, M., Montanari, U. (eds.) Programming 1982. LNCS, vol. 137, pp. 337–351. Springer, Heidelberg (1982). https://doi.org/10.1007/3-540-11494-7_22

[TV12] Tavolato, P., Vogt, F.: Integrating formal methods into computer science curricula at a university of applied sciences. In: TLA+ Workshop at the 18th International Symposium on Formal Methods, Proceedings (2012)

[Win00] Wing, J.M.: Invited talk: weaving formal methods into the undergraduate computer science curriculum (extended abstract). In: Rus, T. (ed.) AMAST 2000. LNCS, vol. 1816, pp. 2–7. Springer, Heidelberg (2000). https://doi.org/10.1007/3-540-45499-3_2

[Zin08] Zingaro, D.: Another approach for resisting student resistance to formal methods. ACM SIGCSE Bull. **40**(4), 56–57 (2008)

Author Index

Aceto, Luca   1

Bordis, Tabea   101

Ettinger, Ran   84

Grätz, Lukas   43
Güdemann, Matthias   18

Hundeshagen, Norbert   35

Ingólfsdóttir, Anna   1

Jacobsen, Frederik Krogsdal   117

Kamburjan, Eduard   43
Körner, Philipp   60
Krings, Sebastian   60

Lange, Martin   35
Lestingi, Livia   75

Runge, Tobias   101

Schaefer, Ina   101

Thüm, Thomas   101

Villadsen, Jørgen   117

Yatapanage, Nisansala   133

Printed in the United States
by Baker & Taylor Publisher Services